Einstein versus Bohr

D. Sachs

Einstein versus Bohr

The Continuing Controversies in Physics

MENDEL SACHS

Department of Physics and Astronomy
State University of New York at Buffalo

Open Court
La Salle, Illinois

First printing 1988.

Printed and bound in the United States of America.

Library of Congress Cataloging-in-Publication Data

Sachs, Mendel.
Einstein versus Bohr : the continuing controversies in physics / Mendel Sachs.
p. cm.
Bibliography: p.
Includes index.
ISBN 0-8126-9064-8 : $32.95. ISBN 0-8126-9065-6 (pbk.) : $15.95
1. Physics—History. 2. Relativity (Physics) 3. Quantum theory. 4. Einstein, Albert, 1879-1955. 5. Bohr, Niels Henrik David, 1855-1962. I. Title.
QC7.S2 1988
530.1—dc19

Dedicated to the memory of my mother
FLORENCE SACHS
for love and encouragement during my formative years

The illustrations for the cover, the frontispiece and figures 1.3, 5.3, 5.4, 5.5, 6.1, 8.1, 8.4, and 8.5 are by Daniel Sachs.

I WANT TO KNOW HOW GOD CREATED THE WORLD.
I AM NOT INTERESTED IN THIS OR THAT PHENOMENON,
IN THE SPECTRUM OF THIS OR THAT ELEMENT;
I WANT TO KNOW HIS THOUGHTS; THE REST
ARE DETAILS.

Albert Einstein[1]

INDEED THE FINITE INTERACTION BETWEEN OBJECT AND
MEASURING AGENCIES CONDITIONED BY THE VERY
EXISTENCE OF THE QUANTUM OF ACTION ENTAILS . . .
THE NECESSITY OF A FINAL RENUNCIATION OF THE
CLASSICAL IDEA OF CAUSALITY AND A RADICAL REVISION
OF OUR ATTITUDE TOWARD THE PROBLEM OF PHYSICAL
REALITY.

Niels Bohr[2]

Table of Contents

List of Illustrations

Foreword

In any introduction to modern philosophy in general or to the theory of knowledge in particular, two candidates for the foundations of science are presented: the *a priori* and the *a posteriori* (meaning the before and the after, meaning before experience and after experience, the deductive and the inductive, the axiomatic and the empirical). They lead respectively to the deductive and the inductive styles of presentation. And philosophers traditionally debate the pros and cons of each of the two proposed views. Now how does one debate pros and cons, deductively or inductively? Indeed, in most presentations, whether deductive or inductive, including even Euclid's *Elements,* one comes to a point where our authors leave the style of their choice and break into debate. Debate, then, has its own style. It is called hypothetic-deductive or, more generally, the dialectical or the argumentative, or still more generally, the problem-oriented.

Most books follow the inductive or the deductive style, and more often than not, a mixture of these. Few authors employ the dialectic style, so very useful at all junctions of scientific revolutions, beginning with Galileo's *Dialogues* and ending with the problem-oriented works of Einstein and Schrödinger.

Sachs has made a valid observation: most popular science presentations avoid problems, or present problems only after they have been solved to the satisfaction of all concerned. Personally, I think there is room for a broader complaint. Often individuals who have graduated with flying colors prove to be utter failures as graduate students, or have severe emotional and intellectual breakdowns out of which they may emerge entirely transformed in order to be able to continue as graduate students. This suggests that what is required of a successful undergraduate does not suffice to qualify one as a research student. This is what Sachs views as an insult to the intelligence of that student. And how right he is.

It is traditional within science to have a tidy shop window and a messy workshop. The individual who graduates well as a science student but is not qualified, or not yet qualified, to work in the scientific research workshop is one who has been thoroughly convinced that science is orderly—inductive, deductive or both—but not the mess of the problem-oriented workshop. The tradition of equating science with the shop window is understandable; it began in the days when science faced hostility and needed strong arguments to recruit novices dedicated to its cause. It is no doubt paradoxical to invite people to the workshop by pretending that science is all shop window. But the paradox is soluble in practice. The non-scientific world outside is impressed by success. The impressed outsider is slowly drawn in and invited to see that success is the outcome of challenge, is challenged to visit the workshop and at times is challenged by some item noticed only in the workshop, and then the visitor settles down in the workshop.

In truth, however, the division between shop window and workshop is an artifact; in reality each item is double-faced, one face successful, one problematic. These days every really challengeable individual must recognize that, and the obstacle to this recognition is the fear that when the shop window is destroyed, the ammunition science has against its enemies is lost. Perhaps. But then, science has no real enemies these days, and anyway, the internal concerns of science are more important than those of its public relations.

For the two sets of concerns do clash. When a newcomer to scientific education is troubled by some weak aspect or rough edge of science as presented to the public—say in the classroom of the undergraduate science class—then the old-timers have two choices: they may insist on discussing the shop window until the newcomers graduate and then allow them a view of the workshop; or they may admit openly that science is all workshop and that every shop window item is also a workshop item. Here, Mendel Sachs's observation signifies that the concealment of the workshop is an insult.

It is a simple empirical fact that a problem-oriented presentation is much easier to comprehend than either a deductive or an inductive one. Presenters who are frank about the shortcomings of the material they present make better sense to their public and are able to direct the public's attention to where they think the action is, while postponing, if need be, discussion of some other problematic aspects of the materials they present. Doing this acknowledges that we cannot solve all problems at once; we always have to postpone attending to some problems while we attend to other problems.

When I was a student of physics I was troubled by the difficulties presented and aired by Sachs in this book. Both the physicists and the philosophers of science to whom I confessed my troubles—as clearly as I could—showed me hostility rather than sympathy. I had earlier experienced the same from my religious teachers, so that I was not crushed by the hostility, but I was discouraged from pursuing my scientific interests. This book has returned me to those days and reminded me of the tremendous joys I experienced then, reading Einstein and Schrödinger and meeting Karl Popper and Alfred Landé. All four expressed, one way or another, the same sentiment as Sachs: feeling difficulties about current ideas should be encouraged, not discouraged.

Perhaps I am biased as a result. For I share the view of Landé and Popper about a major problem in contemporary physics. Most physicists today are indeterminists and subjectivists. I will not explain these terms here, since Sachs does that extremely well in this book. The minority, including Einstein, Schrödinger, Bohm, and Sachs, are determinists and, of course, objectivists. Can one consistently be an objectivist and an indeterminist? Landé and Popper say yes. And I hope that they are right. But the most important thing is not so much the correctness of a view but its rationality. For all the above views are well-known, and each of them has many adherents, some learned and some ignorant. Learned adherents of an erroneous view are preferable to ignorant adherents of a view which may be right, yet which they adhere to not because it is right but simply because they are ignorant and because of some arbitrary choice or some chance in their lives. The learned who are in error may learn more and improve, but the ignorant hold their views for no good reason and then may change them for reasons which are no better.

The problem-oriented approach is thus the best. And when we come to contemporary physics it is essential. For, it is essential to present competing solutions side by side and compare them. Using books each of which presents its author's own ideas will not do, since the slightest variations in meanings of key terms may cause the greatest difficulties. At times, as Sachs shows, authors write at cross-purposes because terms such as the term 'hidden variable' have diverse meanings in diverse contexts.

I do not mean to say that Sachs adjudicates correctly even concerning accepted terminology. Even his discussion of hidden variables may be criticized on account of his strict use of the term; I cannot say. Yet at least he is clear. Concerning hidden variables there is the celebrated proof—von Neumann's proof—of their exclusion from quantum theory. Every time I raised a difficulty with quantum theory I was told to read von Neumann's book. His proof and his book are repeatedly mentioned in the vast literature on the trouble with quantum theory. There is never reference to a page number, or even to a chapter number, where the proof is given. I confess to having studied the book very carefully both by myself and together with two famous students of the matter, Karl Popper and Paul Feyerabend. I still remember the long evenings the three of us spent together in an effort to find the proof in that book. At least for myself I can say that I failed. I challenged a few scholars to tell me where that proof is or to restate it to me. One of them, Abner Shimony, whose contribution to the study of hidden variables is by now so very justly famous, once conceded to me in a conversation that he had attempted to meet my challenge and failed, and then took it back in another conversation, in which he claimed that his concession was due to my having confused him. Well, all is well that ends well.

Sachs offers a proof which is clear, easy to follow and possibly correct. This will force the profession, if anything will, to consider this book seriously as offering an important clarification of a major item in a classical dispute. And those who find fault with Sachs's presentation I now challenge to publish a criticism and preferably also a correction. This is the road to progress.

I have mentioned only two items on which this book is usefully informative: determinism versus subjectivism and von Neumann's theorem. But the book is full of clarifications, presented by the use of the problem-oriented or dialectical style. It is a fascinating study by any reasonable standard, but especially the standard of the problem-oriented dialectician; it deserves careful scrutiny and, of course, correction. It will help keep the debate moving and help to destroy the barrier between shop window and workshop—which happens to be the same barrier as that between by-stander and active researcher. In this book, as in those of Einstein, Schrödinger, Landé, and Popper, the reader can stand close to the anvil and see the sparks flying while the workers forge their tools.

JOSEPH AGASSI

Tel-Aviv University and
York University, Toronto

Acknowledgements

The final corrections to the typescript of this book, and the writing of the last section of the Postscript, 'Implications of Holism: From Tao to Relativity Physics', were completed while I was on sabbatical leave in the 1986–87 academic year. I wish to thank my hosts: Physics Department, University of Canterbury, New Zealand; School of Mathematics, University of New South Wales, Australia; and the Edelstein Center for the History and Philosophy of Science, Technology and Medicine, Hebrew University of Jerusalem, Israel, for their kind hospitalities. I am especially grateful to the Manuscripts and Archives Division of the Jewish National and University Library, at the Hebrew University of Jerusalem, for allowing me access to the Einstein Archives.

Finally, I wish to thank my home institution, State University of New York at Buffalo, for granting me this 1986–87 sabbatical leave of absence—a period that was most conducive to completing this work.

MENDEL SACHS

Department of Physics and Astronomy
State University of New York at Buffalo

Introduction

The primary focus of this book is upon the main philosophical ideas and debates of the two major developments in twentieth-century physics—the quantum theory and the theory of relativity—as proposed explanations of the nature of matter.

It is a lesson of history that we should never accept a *scientific truth* as a *final truth*. Because of our human limitations, and the interconnectedness of all aspects of the physical universe, human beings will continue to be mistaken about what is true and what is not true in science. The search for scientific truth, which in my view is the main purpose of science and philosophy, must entail a continual struggle between conflicting ideas. This is especially so in regard to the major developments of contemporary physics.

It appears to me as a professional physicist that the ideas of science in each period did not appear suddenly, totally disconnected from preceding developments in the history of science. I believe that strands of truth about the physical world do persist throughout all of the so-called 'revolutions' in science,[3] and that real progress is evolutionary rather than revolutionary. It is the continuation of these strands of truth through the different periods of history of science that characterizes actual progress in our understanding of the physical universe. Of course, history does reveal that changes in scientific ideas often occur over short periods of time. Such rapid evolutionary change, though still connected with some of the ideas of the past, then gives the illusion of a genuine revolution of ideas, a complete break with the past. But a closer look reveals that it is indeed evolutionary, after all. It is in this sense of change that I refer below to 'revolution' in science.

There have been many books for lay readers on the notions that underlie the quantum and relativity theories. It may rightly be asked: why another? The answer is that all the literature (I know of) on this subject fails to emphasize a point essential for a true understanding of the history of physics. It is that *progress in science entails conflicts of ideas.* In our time, we see a fundamental conflict between the underpinnings of each of the current explanations of the behavior of elementary matter—the quantum and relativity theories—such that both theories, under the single umbrella of a 'relativistic quantum field theory', yield an unsatisfactory description, from the viewpoints of both logical and mathematical consistency.

This is a fascinating and unique development in the history of science. In the past only one 'revolution' in science occurred at a time. But in the twentieth century two dichotomous 'revolutions' happened simultaneously. What is exciting here is that, on the one hand, each of these theories requires an incorporation of the other for its completion, but on the other hand, each of these theories is logically (and mathematically) incompatible with the other.[4] Thus it seems that with the birth of the quantum and relativity theories there were sown the seeds of a third 'revolution' in science.

But what direction might this new change take? It will be suggested in this book that a reasonable path may be toward a proper generalization of either the quantum theory or the theory of relativity, with an incorporation of the formal expression of the abandoned theory as a particular approximation for the accepted theory, within the limits where the abandoned theory had been empirically successful.

The primary purpose of the presentation of this book is to explain to the general reader precisely what the concepts that underlie the quantum theory and those that underlie the theory of relativity are all about. **I provide some math, but my entire argument can be followed without it.** I hope to reveal in a non-mathematical way the primary conflicts that arise in the attempt to fuse the quantum and relativity theories, as underlying explanations of the behavior of elementary matter, in a single theory. Discussion will then be given to the problem of generalizing one of these theories at the expense of the other, in order to steer the way toward a possible resolution of the present difficulty. Perhaps, then, the reader may appeal to his or her own intuition about which approach is best for progress in our understanding of the physical universe. At least the reader may be brought to the forefront of our *conceptual knowledge* of the behavior of elementary matter—that is, the explanation that may underlie its description.

Recall Einstein's remark that "God may be subtle, but He is not malicious." What I believe he meant by this was that the laws of nature are *conceptually simple.* This idea does not imply, however, that conceptually simple laws have an equally simple mathematical expression. For in the context of science, mathematics, per se, is only a language whose purpose it is to facilitate as precise as possible an expression of the concepts. But it is a lesson of the history of science that the pace of progress in the development of the mathematical language of science has never kept up with the pace of progress in our discoveries of scientific concepts. Our primary concern in this book will be with the concepts of modern physics and their intrinsic controversies, not with mathematical discussions of them.

There is a widespread confusion between conceptual and mathematical simplicity, among those non-specialists who study science as well as among professionals in the field. On this point, it has always been interesting to me that the theory of relativity is extremely simple conceptually, but quite complicated from the point of view of its mathematical expression (as a field theory of matter in general relativity). On the other hand, the second major development of contemporary physics, the quantum theory, is mathematically simple and conceptually difficult. According to Einstein's principle of simplicity—the idea that conceptually simple theories in physics are more likely to be true than conceptually complicated theories—the theory of relativity is more likely to turn out to be true in the long run as an underlying theory of elementary matter than are the ideas of quantum mechanics. This was, in part, at the root of Einstein's disagreement with Bohr and the Copenhagen school on the validity of quantum mechanics as a truly fundamental theory of matter. I shall highlight this controversy in the ensuing chapters of this book.

Some authors of books on concepts of physics for lay readers believe that professional physicists should not expose the innocent reader to the actual conflicts that truly exist in contemporary physics. They feel that the lay reader as well as the serious student of physics can only be enthused by the ideas of science if they are presented as a *fait accompli*—that is, with everything already worked out and explained and neatly presented on a silver platter, without any flaws. I do not concur with this opinion; it insults the reader's intelligence. For such a view, besides being false in claiming that there are no persisting conflicts, has no adventure in it. It is as static as an airline schedule.

Rather, the full story about the present state of ideas in physics, including its persistent conflicts and the lack of exact knowledge about future paths toward understanding the material world, is exciting not only to the professional physicist whose research has to do with the search for the secrets of the universe, but should also be stimulating to lay students of science, as it truly captures the sense of excitement, adventure, and mystery that go with any genuinely intellectual pursuit, that is, pursuit of new ideas. It is an approach that reveals the true essence of scientific discovery and the beauty it unfolds. Above all, such a presentation is not boring.

It is in this spirit of the adventure of scientific exploration that the Medieval philosopher and theologian, Moses Maimonides, wrote the following prayer for the scientists of his day:[5]

> Grant me strength, time, and opportunity to correct what I have acquired, always to extend its domain; for knowledge is immense and the spirit of man can extend infinitely to enrich itself daily with new requirements. Today he can discover his errors of yesterday and tomorrow he may obtain a new light on what he thinks himself sure of today.

One

The Ideas of Classical Physics

To gain some understanding of the concepts of twentieth-century physics it is important to first look into some of their important precursors. In this chapter, we will discuss the sixteenth- and seventeenth-century ideas of Galileo and Newton, the approach that is conventionally referred to as 'classical physics'. We will emphasize the method of pursuing scientific knowledge, especially instigated by Galileo (1564–1642), as well as the physical ideas proposed in the classical period that were seminal for the evolution of current ideas in physics. Then, in chapter two, we will discuss the ideas of nineteenth-century physics that served as an important conceptual bridge between 'classical physics' and the contemporary ideas of the quantum and relativity theories, as basic explanations for the nature of matter.

Galileo's method of investigation involved an interplay between theoretical and experimental pursuits.[6] The idea was first to propose particular concepts by means of 'thought experiments'. These are logical constructs about ideal experiments, experiments that are impossible to carry out in practice because they rely on ideal situations, unachievable in reality. After deducing logical conclusions from these proposals, one then applies them to real experiments, making precise predictions about their outcomes, in qualitative as well as quantitative terms. By carrying out the real experiment, Galileo then compared its numerical results with the theoretical predictions. The idea was that as long as the comparison was favorable, one could claim that the theoretical starting points—the hypotheses of a given theory—had some *scientific truth* in them. Thus, this method of approaching scientific truth is hypothetico-deductive, tied to experimental verification.

It is important to note, with this method of investigation, that 'scientific truth' may only be thought of as a provisional truth that is contingent on nature. This is in contrast with an 'analytic truth' such as the logical deduction:

If A implies B and if B implies C then A implies C

Based on the definition of 'implies' and the rules of a particular sort of logic, the latter conclusion is said to be 'logically necessary'—because there can be no other conclusion than this one.

But a 'scientific conclusion' is based on some initial assertions about the world which may or may not stand up under all possible observational tests and logical tests of consistency with other scientific assertions. That is to say, the axiomatic starting point of a scientific theory about some particular phenomenon *is not necessarily true;* it is rather contingent on the ways of the natural world. As our knowledge progresses and as our methods of experimentation improve in accuracy and variation, we are often faced with inconsistencies with things we felt to be true until that time. The scientist is then forced to alter (or replace altogether) ideas he had previously believed to be scientifically true.

On the other hand, a logically necessary truth refers to a 'convention', rather than to nature itself. For example, with the convention of the real number system and its invented logic (arithmetic), it is necessarily true that $2 + 3 = 5$. But this conclusion may not be true if we choose to use a different logic. To explicate this further, note that one of the rules that leads to this arithmetic equality is the definition of a number in correspondence with an interval along a straight line. The collection of all such numbers forms an 'open set'. Suppose, however, that we choose to define our number system in terms of a closed set, such as the intervals on the circumference of a circle. If we think of the circle as a four-element set, it may then follow that $2 + 3 = 1$ (see figure 1.1).[7]

The idea that these arithmetic 'truths' are relative to a

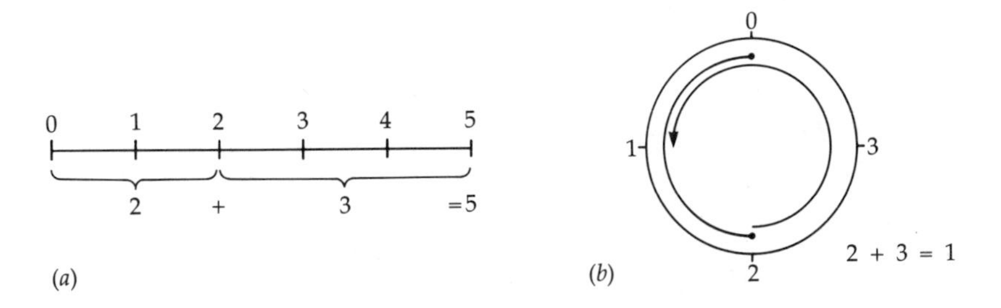

Figure 1.1. The Convention of Adding Numbers. *(a) 2 + 3 = 5 is obtained by superposing the unit interval twice along a straight line (two marks) and then three more times—yielding the mark called '5'. (b) Superposing the unit interval twice and then three times on a 4-interval circle yields the mark called '1'—thus, 2 + 3 = 1.*

particular logic *that we invent* means that they are not 'scientific truths'. This is because the hypotheses about scientific explanations of phenomena relate to the way we suppose the real world to be, independent of our particular participation in it. And we could easily be mistaken about what it is that is basically true about the real world, as we probe it with our theoretical and experimental investigations. On the other hand, the axioms of mathematics are not restrained by the way the real world is. They are pure inventions of our minds, similar to the invention of ordinary language, and not subject to any tests based on the outside world that we probe in science.

What I am saying here is something that Galileo also stressed in his writings—that mathematics, *per se,* is not science. It is, rather, a set of rules that we invent for the express purpose of providing a language with which to represent the laws of matter, and to check their implications in a precise way.

Galileo's Principle of Inertia

An important example of a 'thought experiment' constructed by Galileo is the one he devised in order to arrive at his *principle of inertia.* This is the assertion that if a body should move unimpeded, or if it should be at rest, it would continue in this state of motion *forever,* unless it should be compelled to change this state of motion by some external agent. Thus, Galileo concluded something that no one in his day could have imagined—that no external force is required to maintain a body's *natural* motion, so long as this motion is 'inertial' (which we now know to be in a straight line at constant speed). An exception, in Galileo's day, was Descartes, who also maintained that a natural 'intrinsic' feature of matter is its constant motion. According to Galileo's *principle of inertia,* then, the concept of force has to do with the cause of a *change* of inertial motion (a change of constant speed in a straight line). That is, 'force' (which wasn't described explicitly until Newton, a generation later) is defined as the cause of an accelerating motion.

The way that Galileo concluded his principle of inertia was by first asserting that when a body falls freely toward the Earth it is always directed toward its center. His thought experiment was then to consider a body sliding down an inclined, frictionless plane. He argued that since the body would fall freely if the plane were not present, along a path that is a radius of the Earth, however it may be restrained by an inclined plane, the force that makes it fall must be directed toward the center of the Earth. Thus he concluded that if a body should be held at rest at the top of an inclined plane, h meters vertically above the bottom of the plane (i.e. h is directed along a radius of the Earth), then after it would be released it would slide down the plane with a speed acquired at the bottom that must depend on the magnitude of h. (see figure 1.2). His next step was to suppose that at the bottom of the inclined plane the body might face another upward inclined plane: How far might it go before stopping? Galileo answered that it must proceed up the second inclined plane until it reaches the vertical height h that it originally was released from (see figure 1.2). This is because the top of the plane (corresponding to the vertical height h) corresponds to the state of rest, while the point at the bottom of the plane corresponds to the state of maximum speed.

Galileo then asked further, suppose that the angle of inclination of the second plane is less than that of the first plane. Then it would have to go further before it would reach the vertical height $h,$ and thus it would take more time to stop than the time taken to move from the top of the first plane to the bottom. If the angle of inclination of the second, upward

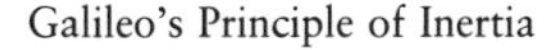

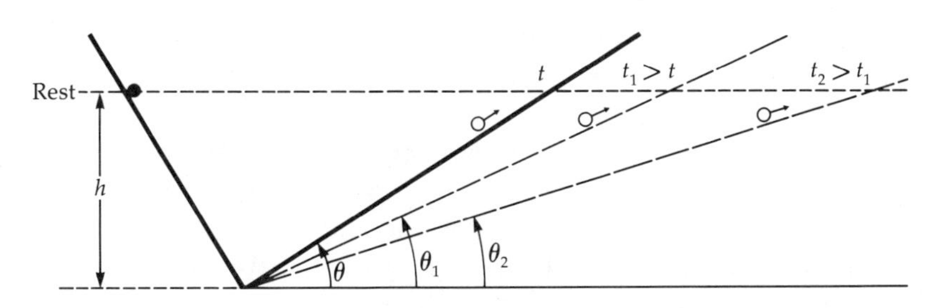

Figure 1.2. Galileo's Principle of Inertia from a Thought Experiment. *A body at rest at the vertical height h is released on a frictionless plane, achieving maximum speed at the bottom. It then proceeds up a second plane, slowing down and then stopping only when it reaches the vertical height h. As $\theta \rightarrow O$, the time taken to come to rest, $t \rightarrow \infty$. Thus, for the case $\theta = O$, the body would move* forever *at the constant speed it had acquired at the bottom of the first inclined plane.*

plane should then be continuously decreased, Galileo argued, it should take a correspondingly longer time before it would come to rest. The limiting case is when the angle of inclination of the second plane becomes zero; in such a case the body would never come to rest since it would never reach the vertical height *h*. Galileo here suggested that the *reason* a body in motion on an upward slope eventually stops is that there is a *part* of the total effect of the Earth on this body (the force of gravity as Newton called it a generation later) that acts to slow it down, that part that depends on the angle of inclination of the plane. In the limiting case, when the angle of inclination of the upward plane is zero, there is no longer any upward motion (toward the vertical height *h*) and thus there is no force to impede the motion of the body. It should then continue to move along the horizontal plane (with the speed it had acquired at the bottom of the first, downward plane) *forever*—so long as there is no impeding force from the outside.[8]

In actual practice there is always a force imposed on the body's motion along its direction of motion due to the friction between the surfaces of the moving body and the plane it moves along. But Galileo's 'thought experiment' was ideal in assuming frictionless planes. His conclusion was then based on a non-observable idealization. Still, it led to further conclusions that are indeed observable in real experiments.

There has been some controversy among historians of

science as to what Galileo meant by the term 'horizontal motion'. Did he mean by this, in his statement of his principle of inertia, that the natural motion of the body must be along the horizon of the Earth, and therefore a circular motion, or does his 'horizontal' refer to a straight path? The majority have agreed that Galileo meant to argue that the natural motion of the unimpeded body must be a circular path, parallel to the circumference of the Earth, and thus not a geodesic (A 'geodesic' is defined as the path of shortest distance between any two of its points. In a Euclidean space such a path is a straight line.) The significance of the geodesic path in the context of Galileo's thought experiment is that it corresponds (in our current vernacular) to the path of minimal energy—it would take external force to remove the body from the geodesic path and change it to some other shaped path. Thus it seems that the 'natural path' according to Galileo's logical exercise is a straight path.

Suppose that we say that instead of a sphere, the Earth is a uniform, infinite two-dimensional plane. Then the 'horizontal' would be the straight line 'natural' path. If the Earth were a spheroid then the 'natural' path of the body would be an ellipse, and so forth. In each of these cases, the 'natural' path is a curve that is parallel to the surface of the body it moves on. But in fact there is only one 'natural' path for an *unimpeded* moving body—it cannot be that a straight line and a circle and an ellipse and so forth are all 'natural'. If there is a unique natural path for an unimpeded body, what is it? To answer this question I suggest that we consider any of these differently shaped possible attracting bodies, and take the limit as their mass densities $\rho \to 0$, keeping their total masses constant. All of these examples, except for the uniform plane, have finite radii of curvature R_i. The limit of $\rho \to 0$ with constant mass then corresponds to an infinite limit for the radii of curvature, $R_i \to \infty$. That is, the limit that yields the *natural path* of the unimpeded 'test body' in motion corresponds to a locus of points with 'no curvature'. Such a path in a Euclidean space is a straight line. (The 'curvature' depends on $1/R_i$, thus 'no curvature' corresponds to a plane rather than any sort of curved surface.)

From this argument I must conclude that in Galileo's original thought experiment (which led him to his *principle of inertia*) the natural path of the unimpeded moving body is indeed a straight line geodesic. This means that *in a vacuum* such would be the natural path of the body, since, in principle, only in a vacuum would the body remain unimpeded in its motion.

To demonstrate a practical example of application of Galileo's principle of inertia, consider a boy, whirling a chestnut at the end of a string, circularly. In this case, the motion of the body is not in a straight line, even though it does have a constant speed at each point along the circumference of the circular path. Thus a force must exist to cause the change of orientation of the body, giving rise to its 'acceleration'. The force provided, of course, is the tension in the string, and its effect is the centripetal acceleration directed toward the center of the circular path. Now if the string should suddenly break, the tension would 'turn off' and the chesnut would fly off in a straight line path, with the speed and direction that it had at the instant that the string broke, in accordance with Galileo's principle of inertia. Galileo then predicted the precise shape of this ensuing path, how long the chestnut would take to reach the ground and so forth, all features of the motion testable in experimentation, both qualitatively and quantitatively.

The former example of a chestnut whirling at the end of a string in a circular path is analogous to planetary motion, except that force is provided not by tension in the string, but by the Sun's gravitational pull; this maintains the planet in its fixed orbit, which in this case is an elliptical rather than a circular path. But if the Sun should suddenly extinguish, the planet would fly off in a straight line path, as the chestnut did, proceeding in a direction that would be tangential to its orbital position when the catastrophe occurred, and with the tangential speed it had at that time.

In the analysis of the preceding example in which a boy rotates a chestnut in a vertical plane at the end of a string near the Earth's surface, the physical assumption that would have been made by Galileo is that the acceleration of a freely falling body is independent of the mass or size of the chestnut or that of the Earth. One generation later, Isaac Newton discovered that this acceleration does indeed depend on the parameters of the Earth—its mass and size. He discovered that the *reason* the chestnut is attracted to the Earth—the reason it has weight—is the same reason that the Earth is attracted to the sun, the moon attracted to the Earth, and so forth. His 'law of universal gravitation' then implied that the acceleration of a body in free fall toward the Earth is not constant, but rather depends on the mass of the Earth and the inverse square of the distance from the falling body to the center of the Earth. Thus Newton concluded that the weight of a body on the Earth depends on these Earth-dependent parameters, rather than being an intrinsic feature of the falling body itself. It implied, for example, that the 'weight' of a body on the moon would be different than its weight on Earth because of the different size and mass of the Moon (a weight and the acceleration of a falling body are reduced to 1/6 of their amounts when on the Moon). Galileo had no suspicion of this, and maintained his law, $g = constant$.

Nevertheless, for Galileo to see the error in his assumption from his empirical studies he would have had to do his experiments on motion, say with his inclined planes, very far from the Earth's surface, or else improve the resolution in his instrumentation so as to reduce the error in his measurements by many orders of magnitude. Because he had no reason to believe that his theory was not true, he did not even attempt to do this. But, for argument's sake, if he had discovered this law, as Newton did a generation later, Galileo would have realized that his earlier expression for the physical situation was at least a very good mathematical approximation for the true theory, in accordance with Newton's law of universal gravitation. Similarly, it was discovered in the twentieth century that Newton's theory had to be replaced by a totally different theory to explain the gravitational phenomenon, a theory which emerged from Einstein's theory of general relativity. But the latter superseding theory does have a mathematical approximation in a particular limit that corresponds with Newton's formal expression of the law of gravity, where the latter has been successful.

We find this *evolution* of one physical theory into another throughout the history of science, with the newer theory always having a mathematical expression that may be approximated by the exact expression of the rejected theory, in the domain where the latter was successful, even though the basic ideas of the earlier theory are in part (sometimes totally) replaced. This sort of merging of an older theory in science into a newer one is sometimes referred to as a principle—*the principle of correspondence*—though there is no *a priori* reason why this principle must necessarily always operate.[9] But in this aspect, in the history of science, we see something of the older (superseded) theory in science that persists as our understanding progresses.

The progression of ideas from Galileo to Newton still retained crucial features of Galileo's original ideas. For one thing, Galileo's *principle of inertia* led Newton to an explicit form for his first two laws of motion and forced him to invent calculus for the express purpose of meeting the need for a precise mathematical way of representing variable motion. Let us now continue with a discussion of Newton's ongoing ideas and contributions to the evolution of scientific discovery.

Isaac Newton and Laws of Force

Isaac Newton (1642–1727) was born in the same year that Galileo died. It was his creative genius that led to an expression of Galileo's ideas about variable motion as caused by external forces. Thus, Galileo's principle of inertia was expressed, first, by the relation that we now call *Newton's first law of motion: If no external force acts on a body, then it has no acceleration.* In this case, then, the body must have a constant velocity, as its 'natural' motion. An important special case of constant velocity is the state of rest (relative to an observer).[10]

Now if a body moves at constant velocity, its path is a straight line—a fact readily proven by solving the equation of motion that corresponds to Newton's first law of motion, above. Thus, this law is in precise correspondence with one of the implications of Galileo's *principle of inertia,* assuming, as I have shown above, that Galileo's meaning of 'horizontal' motion is that of 'straight line' motion, as the natural state of unimpeded moving matter.

If an external force should act on a body, then Newton became specific about the precise way in which the speed of that body must change. He asserted that the relation between the applied, external force (the 'cause') and the acceleration of the body that it causes (the 'effect') must be *linear*. That is, doubling the magnitude of the force would double the magnitude of the acceleration, etc. Specifically, if a force F_1 should cause a body to accelerate at the rate a_1 and if a different force F_2 should cause *the same body* to accelerate at the rate a_2, then it was Newton's contention that generally the magnitude of the ratio of forces must be equal to the ratio of the corresponding magnitudes of accelerations,

$$(1) \qquad |\boldsymbol{F}_2|/|\boldsymbol{F}_1| = |\boldsymbol{a}_2|/|\boldsymbol{a}_1|$$

(where the vertical bars denote the magnitudes of the vectors that are the oriented force and acceleration). Thus, equation (1) is an expression of *Newton's second law of motion.*

An equivalent way of expressing this law of motion is to say that the (vector) force that acts on a body is linearly proportional to its (vector) acceleration, that is,

$$(2) \qquad \boldsymbol{F} = \mathrm{m}\boldsymbol{a}$$

where m is the constant of proportionality between the cause *(**F**)* and the effect *(**a**).* It is an intrinsic property of the body that is a measure of its resistance to a change of state of constant motion (or rest). This property is called its 'inertial mass'. For example, with a constant external force applied to a body, a large inertial mass would correspond to a small acceleration or a small mass would correspond to a large acceleration—a tug boat barely moves a very massive steamship from the dock, but with the same force it would accelerate a row boat from the dock very rapidly.

Newton's second law of motion (2) is called a dynamical relation because it prescribes an effect as due to a cause. It doesn't specify the explicit nature of the force (for example, whether it is a gravitational force that depends on the separation of the interacting bodies as $1/r^2$, or whether it is a magnetic force depending on different parameters, and so forth). But this law says that no matter what is the explicit nature of the external force that acts on a body with mass *m*, the acceleration it produces as an effect must be determined by it according to the linear relation given by equation (2).

The form of Newton's second law, given in equation (2), tacitly assumes an atomistic model of matter. For the inertial mass *m* is an *intrinsic property* of a body, that is acted upon by an *external* agent. It was further assumed in Newton's analysis that the force *F* between interacting bodies depends only on their spatial separation. That is, he assumed that the forces exerted by bodies on other bodies, such as the force exerted by the Sun on Earth, depends only on their mutual separation—it is 'action-at-a-distance'. It was indeed disturbing to Newton to have to claim that something that is 'here' can affect something 'over there', without making any bodily contact, no matter how far apart the interacting things may be. For example, if the Sun should suddenly disintegrate, our planet Earth (and the other planets of our solar system) would respond spontaneously, according to the 'action-at-a-distance' model, flying off in all directions along straight line paths, without having had any contact with the Sun. Bothersome as this was to Newton, he said that it 'worked' and that as a scientist, he did not form hypotheses but rather found the formulas that fit the data. Nevertheless, in reading his main works, it is clear that Newton did indeed form hypotheses to underlie the empirical descriptions of natural phenomena. That is, his epistemological outlook was not one of empiricism, as his statements seem to indicate, but rather one of realism.

Mach's Interpretation of Inertial Mass—The Mach Principle

Some 200 years after Newton, Ernst Mach noted that the actual datum that verifies Newton's second law of motion is the relation of ratios (1) rather than the *deduced* equation (2). He commented that equation (2) needn't be interpreted in terms of an intrinsic mass, as Newton did in support of his atomistic model of matter. Mach argued as follows:[11]

Suppose that two different forces should act on *different*

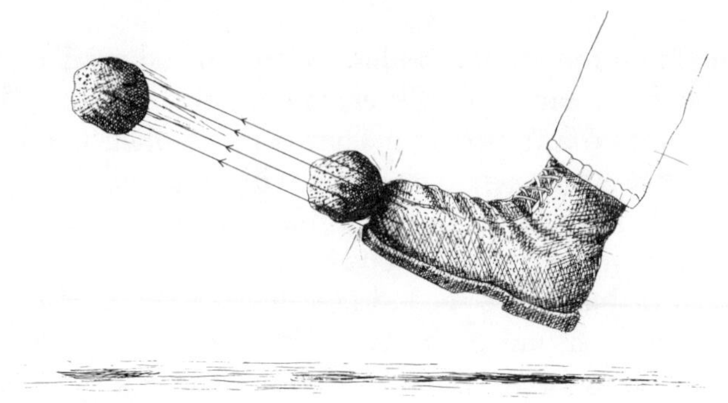

Figure 1.3. Action-at-a-Distance. *When the stone is kicked, does its direct contact with the boot cause it to move, or is its motion due to the action-at-a-distance between the constituent atoms and molecules of the stone and those of the boot?*

bodies (rather than on the same body, as in Newton's analysis), but this time in such a way so as to accelerate each body at the same rate. An example of this would be the force of gravity causing separate bodies to accelerate freely toward Earth at the same rate, when they are near the surface of the Earth. In this case, Newton's second law of motion could be re-expressed in the form:

$$|\boldsymbol{F}_2|/|\boldsymbol{F}_1| = m_2/m_1 \quad (3)$$

Just as Newton arrived at the deduced equation (2) from the empirical relation (1), so the ratio of masses in Mach's equation (3) leads to the deduced relation:

$$m = k|\boldsymbol{F}| \quad (4)$$

Now this mass-force relation may be interpreted quite differently from Newton's atomistic interpretation of his second law of motion. For equation (4) appears to say that the inertial mass of a body is *caused by* the total external force that acts on this body. According to this view, the inertial mass of any quantity of matter is not an intrinsic property of that matter. It is rather an expression of *coupling* between this body and all other bodies that interact with it, that in themselves give rise to the total external force $\boldsymbol{F}$ that acts on that body. In Mach's view, then, inertial mass must have a strictly relativistic connotation and matter must be described, not atomistically, but rather in terms of a *closed system*. This is because the forces that matter exerts on matter generally have infinite range. The implication here is that the inertial mass parameter for any quantity of matter depends on all of the other matter of the universe—since, in principle, the latter gives rise to the total external force that acts on the observed matter that has the inertial mass m.

In our own time, Einstein has called this relativistic interpretation of inertial mass 'The Mach Principle'. It was a view of matter that was very influential in Einstein's thinking, especially in making the transition from the atomistic model of matter that he held in the early stages of his relativity theory (when he discovered 'special relativity') to the later continuum view of matter that emerged with the theory of general relativity. When this holistic idea, originally planted by Mach, was fully exploited in Einstein's theory of relativity, the revolutionary concept emerged that the universe is indeed not a sum of parts, as implied by the atomistic theories, but it is rather a closed system without actually separable parts. This will be discussed in detail in chapter eight.

In epistemology, Mach rejected Einstein's theory of general relativity because he was opposed to any view that entails an underlying reality. That is, Mach's positivistic view was that all that can be meaningful in science must relate directly to physical effects that can be 'sensed'. His view would then have automatically rejected the 'metaphysical' ingredients of Einstein's general relativity theory and its philosophic stand of (abstract) realism. We will return to these important conflicting ideas when we develop more fully the opposing ideas of the quantum and relativity theories of contemporary physics.

Coming back to Newton, two centuries before the contemporary period, it is important to emphasize the particular nature of the force he evoked to explain gravitational phenomena. First, he assumed (with Galileo) that the gravitational force of the Earth is directed along the line of centers between the body with weight and its own center—thus along the radial lines of the spherical Earth. In his generalization, he then believed that all gravitationally interacting bodies interact along their mutual line of centers, with the mutual force of attraction depending inversely on the square of their separation, $1/r^2$. When this form of the force $F(r)$ was inserted into his second law of motion (2), its solutions gave predictions that agreed with all of the known features of planetary motion of his time, particularly the detailed behavior as observed a generation earlier by Tycho Brahe and Johannes Kepler.

It was Kepler who interpreted Brahe's astronomical data in a phenomenological fashion, expressing them in compact form in terms of three regularities that he saw in them, regularities that have been called 'laws of planetary motion': 1) planetary orbits are elliptical, with the center of the Sun at one of the foci of the ellipse; 2) the radial vector locating a planet in its orbit from the Sun's center, sweeps out equal areas of the ellipse in equal times; and 3) the ratio of the square of the period of a planet's motion (the time to complete one cycle) to the cube of its distance from the Sun is a constant, from one planetary orbit to another. One of the stunning successes of Newton's theory of universal gravitation was its accurate prediction of all three of Kepler's laws of planetary motion.[12]

Newton's theory also correctly predicted the pull of the Earth on all terrestrial objects (giving them 'weight'), the ocean tides, and other phenomena within our solar system. As we saw earlier, he found that Galileo's gravitational law, $g =$ constant, was a good mathematical approximation for a different law, near the surface of the Earth. Of course, we are now aware, in the age of space travel, that as we depart from Earth the weight of an object becomes continuously smaller, until it becomes 'weightless'.

In Newton's theory of universal gravitation, 'weightlessness' can occur in two ways. If an astronaut is far out in space, not near any planet or star, he would normally float—not being pulled in any particular direction. He would be 'weightless' because there is no external force acting on his body. However, one may also experience weightlessness near the Earth's surface if there are several external forces acting on his body that cancel each other out. For example, by standing in a vehicle that falls freely toward the Earth (near its surface) one would be accelerating downward at a rate of g **meters/sec²** relative to the Earth; one would then have no weight relative to the floor of the vehicle. This may be seen as follows: Consider at first that the man is standing in a vehicle at rest on the surface of the Earth. The total force on his body in this case is zero, since it is the sum of the force on his body due to the Earth's pull—his weight +mg—plus the force on his body exerted upward by the floor (in preventing him from falling through it!). The latter force is $-mg$ (being directed in the opposite direction from the force of the earth on his body, downward.) The total force is then, in this case,

$$F = ma = +mg - mg = 0$$

implying that the man does not accelerate through the floor of the vehicle.

Suppose now that the vehicle is falling freely toward Earth at the acceleration $a = g$ **m/sec²**. In this case, the total force on his body is:

$$F = ma = mg = F(\text{Earth on man}) + F(\text{floor on man})$$

Since F(Earth on man) in this equation is exactly equal to mg, this cancels the mg on the left side of the equation and we see that in free fall,

$$F(\text{floor on man}) = 0$$

Now according to Newton's third law of motion, asserting that for every force exerted by one body on a second, there is an equal and opposite force exerted by the second body on the first, if follows that the force exerted by the man on the floor must also be zero. That is, if the man would be standing on a spring scale on the floor of the vehicle, while it is in free fall toward Earth, it would register 'zero weight'. That is, the man would then be in a state of weightlessness with respect to the enclosing vehicle.

As we have discussed earlier, it was also implicit in Newton's natural philosophy that matter and the world, generally, are fundamentally *atomistic*. That is, observable matter is assumed to be an assemblage of bits that exchange mutual forces at a distance, thereby causing each other to move in the way that they do. This approach then led many of Newton's followers to the view of 'Newtonian cosmology'. This is the idea that the entire universe—its stars, planets, comets, etc.—are all in stationary orbits, similar to the orbits of the planets within our own solar system. Of course, Newton's theory of universal gravitation had not been checked beyond the (relatively infinitesimal) domain of our solar system. However it was assumed that the law he discovered was a general truth, and therefore that it must necessarily extend to the entire range of the universe.

Such an assumption was incompatible with Newton's own claim that he did not form hypotheses apart from the empirical evidence.

As a general law, other scholars have maintained that Newton's approach in physics applies to other domains, such as the natural laws that govern the behavior of people. This was the *materialistic philosophy* that has been applied by many thinkers after Newton's time to all sorts of problems of the world, outside the domain of physics *per se*. But it was also recognized by other scholars that the materialistic view may be naive in reference to the fundamental nature of human beings and their societies, or even as a last word in regard to inanimate matter itself. The various views of the anthropologists and sociologists throughout the ages, from Newton to the present time, do not all base their explanations for human behavior on the materialistic philosophy. In physics itself, a very prominent deviation from Newtonian materialism occurred in the nineteenth century when Michael Faraday introduced the concept of the continuous field to underlie our understanding of the inanimate, material world from a fundamental standpoint. This view will be discussed in more detail in the next chapter.

Newton's original writings reveal that he did not fully subscribe to the materialistic philosophy. But he did agree with the importance of the notion of generalization in our search for scientific truth. That is, he believed that if a particular set of concepts are true in one domain of physical phenomena, they should also be true in any other domain. Thus he was led to believe that the atomistic concept of matter and the notion of 'action-at-a-distance' must also apply to other domains of physics, such as a basic explanation for optical phenomena.

Newton's Views of the Nature of Light

The conflict between Newton and several of his contemporaries in seventeenth-century Europe on the nature of light had to do with the question of whether light is fundamentally wave-like or particle-like. This conflict about light continued through the nineteenth century, until James Clerk Maxwell discovered that indeed light is fully represented by the wave solutions of the equations of electromagnetism. However, unhappily, this seeming settlement of the 200-year old controversy did not last too long. For in the early part of the twentieth century, light appeared to be schizophrenic—under certain types of experimentation light was seen to be corpuscular in nature, as Newton would have anticipated, but under other sorts of experimentation, light revealed the nature of continuous waves. This puzzle, later named 'wave-particle dualism', led to one of the two revolutions of twentieth-century physics, called 'the quantum theory'. This will be discussed in detail in chapters four through seven. First, however, the present section of this chapter will discuss some of the explicit arguments on both sides of the controversy about light, from Newton's time until now.

Just as Newton deduced that any visible quantity of matter must be made out of discrete 'parts', acting on each other at a distance, so he surmised that light itself is a collection of discrete corpuscles, all travelling along their own trajectories in such a way that it would be possible to explain all of the known optical phenomena.[13]

Newton saw that light travels through different types of material media at different speeds. He then extrapolated to the case where light would propagate through a vacuum. He concluded that this would not be possible and so asserted that there must be an *ether* medium that conducts light. Such a medium must permeate all of space, whether in a dense or a rarefied region of ordinary matter. With the atomistic model, Newton then assumed that this ether medium must be made up of an assemblage of ether atoms which in turn would displace the corpuscles of light as they propagate from one place to another.

What were the optical phenomena that Newton's theory was to explain?

The Reflection of Light

The first phenomenon is the reflection of light, say from a silvered mirror or from the surface of a pool of water. According to Newton, what happens when light is reflected is the following: as the corpuscles of light approach the reflecting material, the material atoms repel them when they come sufficiently close. Thus, the particles of light are different from ordinary 'gravitational' atoms since the latter can only attract each other while the light 'atoms' can be repelled by the material atoms. Newton then claimed that when the atoms of light are sufficiently close to the material atoms of the reflector, the mutually repulsive force exerted on them bounces the light atoms upward. It was then proven that the angle of incidence of the incoming light corpuscles (their angle of approach relative to the perpendicular to the mirror) is equal to their angle of reflection (see figure 1.4). This prediction of Newton's theory, which is in agreement with the data, follows from the law of conservation of momentum—which in turn is a consequence of *Newton's third law of motion*—the assertion that for every

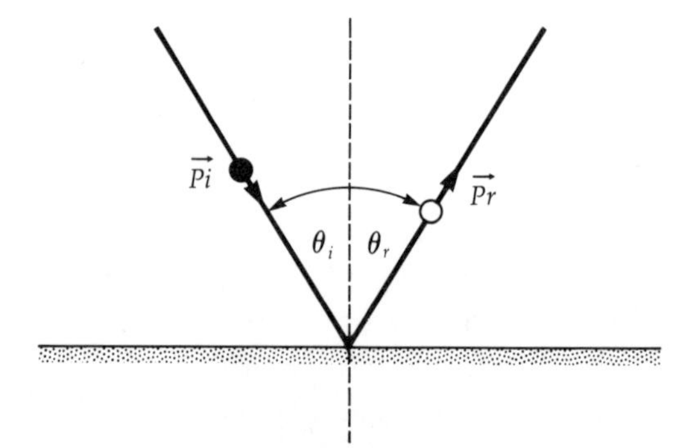

Figure 1.4. The Law of Reflection, $\theta_i = \theta_r$. *If p is the momentum of the light corpuscle and m is its mass, the kinetic energy is equal to $p^2/2m$. If the mirror acts as an immovable wall, the collision of the light with it is 'elastic', and according to: Conservation of momentum: $p_i \sin\theta_i = p_r \sin\theta_r$; according to Conservation of energy: $p_i^2/2m = p_r^2/2m$. Combining these two equations yields the law of reflection, that the angle of incidence θ_i is equal to the angle of reflection θ_r.*

force that one body exerts on a second body, there is an equal and opposite force exerted by the second body on the first.

The Refraction of Light

How does Newton's theory handle the phenomenon of the refraction of light? This is the effect whereby a ray of light is seen to suddenly change its direction of propagation when moving from one medium (say air) into another (say water). It appears to always bend to the perpendicular to the surface of the refracting medium when the latter is more dense than the medium that the light came out of.

One may see this effect in the observation that an object at the bottom of a pond is not really where it appears to be when we view it from the air. The illusion happens because when the light passes from the air into the water to reflect from the observed object, it bends as it propagates into the water and then it bends again after it leaves the surface of the water, propagating toward one's eyes (see figure 1.5).

Newton explained this phenomenon by saying that because the water is more dense than the air it must exert a greater force on the atoms of light than did the atoms of air, thereby causing the light particles to move faster in the denser medium. But he was at a loss to explain the direction of the bending of

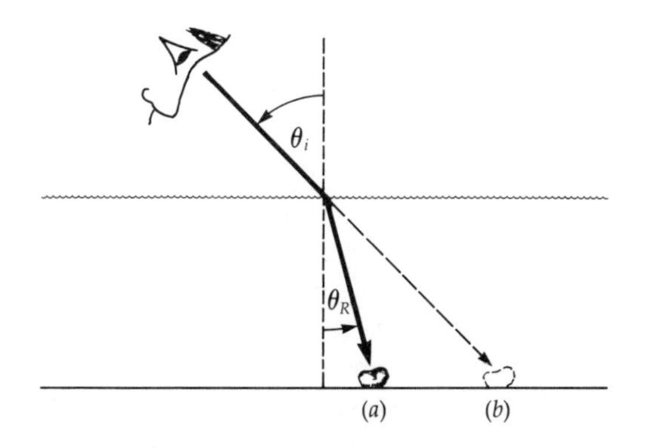

Figure 1.5. The Refraction of Light. *The object at the bottom of a pond, at (a), seems to the observer to be at (b)—because of the bending of the light as it leaves and enters the air from the water.* $\theta_R < \theta_i$ *whenever the refracting medium is more dense than the medium from which it came.*

the light ray as it entered the more dense medium. That is, why does the light ray bend toward the perpendicular to the surface rather than bending away from it as it enters the more dense medium? Secondly, Newton could not explain the fact that only a part of the propagating light is transmitted into the refracting medium (the water in this example) and a part is reflected from its surface. He had no answers to these questions except to respond that there are 'easy and hard fits' of the ether atoms in the vicinity of the refracting medium, so displacing the light particles in one way or another as they approach the surface of this medium.

The reflection and refraction of light were then understandable to Newton in terms of the discrete trajectories of the corpuscles of light and their interactions with the ether atoms and their (repulsive and attractive) interactions with the ordinary atoms of the reflecting or refracting media. But Newton was also aware that there are optical phenomena that suggest that light is a wave, such as the phenomenon of *interference*. For example, *one wave* plus *one wave* (each having the same amplitude) add up to a wave of any amplitude at all between zero and two times the original amplitude of one of the waves. (see figure 1.6). On the other hand, *one particle* plus *one particle* can only add up to two particles and nothing else.

In addition to this conceptual difference between waves and particles, in regard to their respective rules of combination, there is also the conceptual difference in regard to the discreteness (locality) of particles versus the continuity of waves. That is to say, waves are continuously distributed entities throughout space, while particles are localized to special places—if a particle is here, at some time, it can be nowhere else at that time. On the other hand, a single wave could have a non-zero amplitude everywhere, at any particular time.

An example of optical interference that was known about in Newton's time was the observation that light from the Sun, in falling on a thin film of liquid (say oil) coating a reflecting surface, reveals a set of circular rings, alternately dark and light. Newton felt that he was obligated to explain these 'Newton rings' with his particle theory of light, even though it seemed that their natural explanation was with a wave theory—because

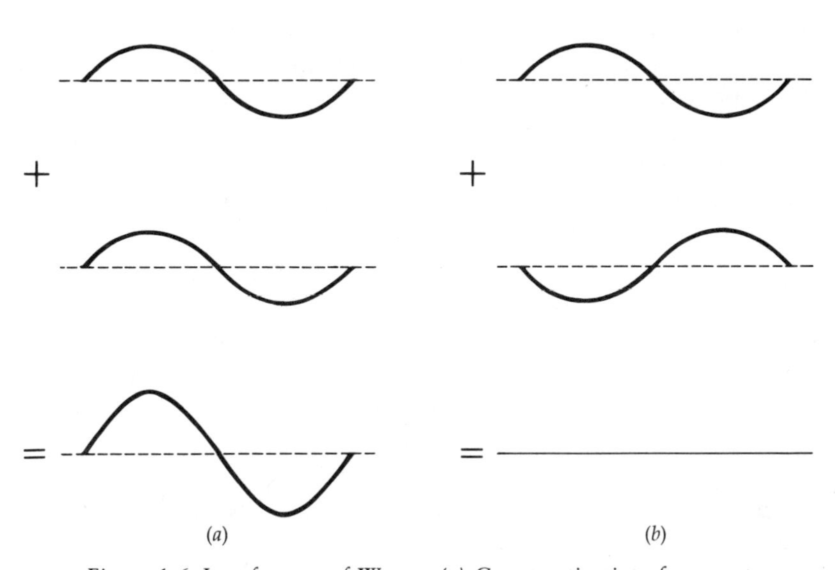

Figure 1.6. Interference of Waves. *(a) Constructive interference: two waves of equal amplitude are in phase. In this case their sum is a wave with twice their amplitude. (b) Destructive interference: two waves are 180° out of phase. In this case their sum yields* no wave. *Altering the relative phases of the two waves arbitrarily correspondingly alters their sum continuously.*

waves naturally combine to produce 'interference fringes' that could relate to these dark and light rings. What Newton said about this was that in these cases there must be a sort of selective refraction taking place, with forces that entail a complicated alternation between repulsion and attraction. This operates between the particles of light and those of ordinary matter as the light moves through the refracting medium at different angles from the perpendicular to its surface.

The Dispersion of Light

The dispersion of sunlight into a spectrum of colors, such as the rainbow, was explained in Newton's theory by asserting that the 'white light' that comes from the Sun is composed of different sorts of corpuscles, each corresponding to a different color (from deep red to deep violet). The idea was that in the ether all of these corpuscles travel at the same speed; but in other (ordinary) material media, such as glass or water vapor, they each travel at a different speed. Thus, if white light should enter a material medium at some initial time, each of the different sorts of corpuscles, corresponding to the different colors, would emerge from the medium at a different time, thereby revealing the entire spectrum of colors. Newton's model then explained the colors of the rainbow in terms of the dispersion of sunlight as it passes through the water droplets during a rain. Nevertheless, the theory did not provide an explanation for the interactions that cause the different light particles to move at different speeds, nor did it explain other features of the rainbow, such as its shape and the fact that it always meets the horizon at the same angle.

The Diffraction of Light

One important physical phenomenon that Newton tried to explain with his optical theory in terms of particles was that of the diffraction of light. This is the appearance of bending around corners that light displays under the proper circumstances. If a beam of light should fall on a screen with a hole in it, and if there should be a second screen on the other side of the first one, some distance beyond it, it would be observed that if the hole in the first screen is sufficiently small, the image of this hole on the second screen would not be sharp. Further, if the hole in the first screen should be continuously decreased in size, the image of the hole on the second screen becomes continuously more diffused. The problem is that if Newton's particle model of light were correct, the image on the second screen of the light coming through the hole in the first one should always be sharp and it should decrease (rather than increase) as the hole size is diminished.

Finally, if one should look closely at the image on the second screen it would be seen that the diffuse pattern of the hole has some structure. There is a central region with maximum intensity of light, opposite the hole in the first screen, but there are also alternate fringes of dark and light rings (corresponding to the regions where the light does and does not fall) continuing outwards with decreasing intensity. This is called a 'diffraction pattern'. The phenomenon was first observed, two centuries before Newton, by Leonardo da Vinci.

It is easily explained by a wave theory of light. But Newton thought he could explain it in terms of purely refractive and reflective forces between the particles of light and the matter of the screens.

There is a feature of light diffraction that Newton did not seem to know about, and does not seem at all explicable with his model. It is most easily described with the Young double slit experiment (see figure 1.7). Suppose that there are two holes instead of one in the first screen, in the preceding example. According to a wave theory of light, as will be discussed further below, if light of a fixed frequency should fall on this screen, there is a prediction of a diffraction pattern on the second screen that has a central maximum intensity just opposite the solid section of the first screen, between the two holes. How could a particle theory of light explain this?

Newton did carry out experimentation on the diffraction of light. He observed the diffraction pattern formed by the scattering of light from a thin hair. Had he seen a part of this diffraction pattern in the geometrical shadow of the hair, he might have concluded that his corpuscular theory of light simply

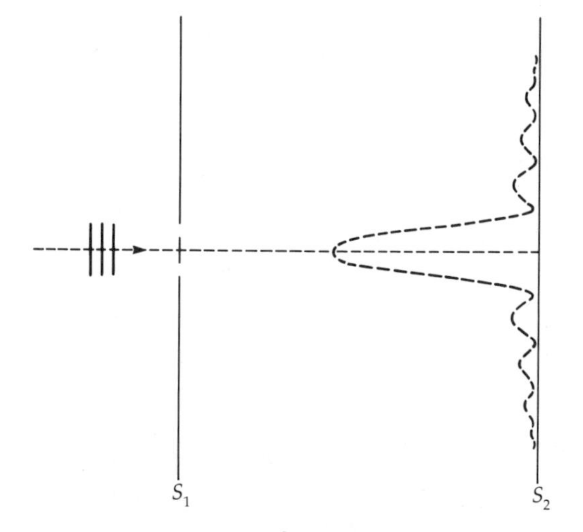

Figure 1.7. Young's Double Slit Experiment. *The diffraction pattern on the second screen, S_2, shows the light intensity after passing through the two slits on screen S_1.*

failed to explain the diffraction phenomenon and he might have then yielded to the wave theory which provides an easy explanation for the phenomenon.

The Early Wave Theory of Light

Some of Newton's contemporaries, notably Hooke (in England) and Huygens (in Holland),[14] believed that light is strictly a wave phenomenon. It was assumed to be a propagating longitudinal vibration of the ether medium, analogous to sound, as a propagation of a wave through a material medium. The 'ray' property of light was seen by Huygens to be a directed 'wave front'. His model was as follows: A light wave spreads out from a point source as a spherically distributed vibratory motion. But then each point on the spherically expanding wave front acts as a new point source for a new spherical wave. Thus, a continuum of spherical waves are produced along a spherical wave front, that originally emanated from a single point source. All of these waves are then said to add up constructively with their phases so adjusted as to give rise to a propagating wave front in the forward direction, and they destructively cancel in the backward direction. It is the forward propagating wave that relates to the 'ray' in geometrical optics. This model is called the 'Huygens principle'.

Huygens' wave theory of light predicted the law of reflection (that the angle of incidence is equal to the angle of reflection); Newton's theory had also done this though it was based on a particle view. But in regard to the refraction of light, Huygens' result was opposite to that of Newton. What he found was that if the angle of incidence (of the light coming from the less dense medium) is greater than the angle of refraction (of the light entering the more dense medium), *as observed,* this implies that the speed of the light in the less dense medium is greater than it is in the more dense medium. It also implies that the maximum speed of light occurs in a vacuum (that is, when there is no material medium at all, except for the ether). This result is the opposite of Newton's prediction, which is that because of the greater force of the material atoms on the light particles in the more dense medium, the light should propagate with greater speed in the more dense medium. Thus, the refraction phenomenon provided a clear-cut test of Newton's corpuscular theory of light against Huygens' wave theory. Unfortunately, experimental procedures had not yet reached the stage in the seventeenth century that would allow such a test to be carried out. Later on, of course, when such experiments were conducted on the speed of light in different media, it was conclusively established that Huygens' conclusion was correct, and that the maximum speed of light does occur in a vacuum.

While Huygens' wave theory of light was able to duplicate all of the optical phenomena that were known in the seventeenth century, so could Newton's theory. The true nature of light then remained controversial for many generations, when further experimentation favored the wave theory. Theoretical studies in the nineteeth century then led to the spectacular discovery that light is nothing more than the propagation of an electromagnetic force between coupled charged bodies. Such propagation, called 'electromagnetic radiation', is a transverse wave, that is, it is a wave that oscillates in a plane perpendicular to its direction of motion. This is in contrast to a longitudinal wave, as Huygens had envisioned for light, which oscillates along the same direction as its motion. It was found that the different frequency solutions of the equations for electromagnetism not only predicted the frequencies associated with the different colors of light (the spectrum of the rainbow) but other frequency solutions corresponded to other physical phenomena—radio waves, X-rays, gamma rays, and so forth all from the same laws of electromagnetism.

As we have indicated earlier, the state of contentment about finally understanding light at the end of the nineteenth century did not last long, for new experimentation in the early years of the twentieth century implied that under particular experimental conditions, light might appear once again as a 'particle', while under other conditions it would still appear as waves. This led to the concept of 'wave-particle dualism', that was to underlie the revolution of quantum theory in twentieth-century physics, to be discussed in several later chapters of this book.

René Descartes (1596–1650)

Returning to the generation of Galileo in the Renaissance period of seventeeth-century Europe, the French scholar, René Descartes, also contributed significantly to the scientific and philosophic thinking of that period which was influential on contemporary views in physics.[15]

The Cartesian view of matter was in one sense atomistic, but in another sense it took the view of continua. Descartes asserted that the only permanent attributes of matter are its *extension* and its *motion*. Other qualities of matter, such as color, hardness, temperature, etc. were assumed in his thinking to be changeable and therefore not related to the *essence* of matter. He argued that since one observes a change in the extension of a quantity of matter—for example, the contraction or expansion of a metal when it is cooled or heated—it must be made out of *parts,* whereby each constituent part of some observed matter changes its location relative to the other parts of that matter *in space*. Thus he concluded that space, *per se,* must be an attribute of matter because extension is in space. However, he argued, a change of form is *in time*. Motion, which he also took to be an attribute of matter, refers to both spatial and temporal intervals—it is a change of a spatial interval with respect to a temporal interval. Because of Descartes' assumption that motion, *per se,* is a permanent attribute of matter, he concluded that motion of matter is not caused by any external agent, though it must be a *constant motion*. However, it then followed that if a quantity of matter's motion would change, this must be due to some external agent. This conclusion is then quite similar to Galileo's *principle of inertia,* as we have discussed in the early part of this chapter, and to Newton's first two laws of motion formulated with his newly invented calculus a generation after Galileo and Descartes.

The Cartesian view of matter then seems to be atomistic because it says that any macroscopic quantity of matter has constituent parts, each of them having extension. On the other hand, the claim that extension, *per se,* is a permanent attribute of matter makes it appear that matter, *per se,* is fundamentally rooted in a continuum. That is, because of the assumption that space is a continuum, and that the points of space are defined as entities necessarily occupied by matter, it follows that matter must be distributed continuously in space, rather than discretely, as the atomistic view would hold.

If matter is indeed continuous in this way, what are the apparent 'things' that we see about us? What are the planets, stars, rocks, trees and people, if they are not separate things? In answering this question from a Cartesian viewpoint, it might be said that matter is fundamentally a continuum, though manifesting itself as 'things', just as the continuous ocean manifests itself in ways that appear as separate things—tidal waves, ocean breakers at the shore, and so forth. But with this continuum view of matter, the universe in reality is a single entity, manifesting itself as apparent 'things', yet an entity that is in fact without parts. It is an everlasting, unchanging universe, though its manifestations are continually changing. This is a view similar to Plato's interpretation of the universe.

If all of this is true, then what does Descartes have in mind when he refers to the 'motion of matter'? That is, if matter truly fills all of space, how can it move? He would have to answer this question by saying that the word 'motion' must be interpreted as *transfer of action.* This would be a transfer between different possible modes of behavior of the single system—just as the ripples *of* a pond (rather than parts *in* it) may interact with each other to give some observable outcome in the nature of the pond as a whole. In this case, one would not interpret the 'ripple' as a truly individuated entity, separable from the pond! That is, one may see a ripple gliding along the surface of the pond and then transfer some of its momentum to another ripple that may cross its path. Even so, the two ripples are not really individual parts, separable from the pond and then viewed as things on their own. Rather, they are both manifestations of a single pond.

From my reading of Descartes, I conclude that in the final analysis, he would agree with a continuum view of the universe and thus disagree with Newton on the atomicity of matter. The 'things' of Descartes' universe are modes of a continuous matter field, filling all of space. What appeared to him as separate things, planets, people or galaxies, are then the distinguishable modes of a continuous matter field, like the vortices of a turbulent fluid.

In summary, it appears to me that Descartes took the philosophic stand of 'abstract realism'. That is to say, to him the primary features of matter that *underlie* its observable behavior are not directly perceivable things, they are rather abstract relations. Such an abstract relation is a vortex of a matter continuum that in particular may underlie some observable behavior, to *explain,* for example, the phases of the moon.

The method of proving the scientific validity of such abstract relations is *hypothetico-deductive.* Just as in Galileo's method of pursuing scientific truth, one first hypothesizes basic ideas about the explanations for the phenomena. Then, by a process of logical and mathematical deduction, one arrives at testable conclusions. Descartes' and Galileo's mode of investigating the truths of the real world was based on the tacit assumption that our ability to reason must be utilized in order to achieve a fundamental understanding of the material world. They were indeed leading spokesmen in this period in the history of ideas, the 'Age of Reason'.

Descartes' Mind–Matter Dichotomy

Though matter seemed in Descartes' opinion to be a continuous whole, there was still one aspect that he felt should be viewed atomistically. This is the aspect of the human mind that shows itself in the thinking process. He addressed the following question: How can one logically reconcile the oneness of the inanimate aspects of the world with the plurality of human minds?[16] He next asked the following questions. What is mind? What does it mean to think? Are the processes that underlie the thinking mechanism truly mechanistic, governed by the same natural laws as those that explain the behavior of inanimate matter? His answer to this question was in the negative. Mind and matter are separate entities, not at all based on the same sorts of underlying natural laws.

If this was so, Descartes continued, then how should one explain the seemingly obvious interaction between mind and matter? For example, 1) if I should see a rose (matter), then 2) I might think about picking it up (mind). This thought then acts on my body to cause me to pick up the rose (matter) and then to enjoy its fragrance (mind). It was then Descartes' problem to explain this apparent interaction between mind and matter while still maintaining a rejection of the idea that the mechanistic laws that govern matter also govern mind.

His initial reply to this question was to assert that mind and matter are in one-to-one correspondence, analogous to two very distant stars, moving at the same speed along parallel orbits. While each segment of orbit for one star corresponds to a segment of orbit for the other star, the stars are not actually interacting. That is, their paths correspond but they do not physically influence each other. He thought that the relation between mind and matter was of the same kind.

In his later period, Descartes abandoned this view as perhaps too *ad hoc* and illogical in view of the holistic approach he maintained for the material world. He then contended that there must be some (yet undiscovered) gland in the human body that couples physical action and the thinking action of the mind. The pineal gland was later chosen as a likely candidate, since at that time there was no medical evidence that this gland had any other function.

Baruch Spinoza (1632–1677)

One of Descartes' contemporaries in philosophy who did not accept the mind-matter dichotomy as a real problem was Spinoza.[17] While Descartes saw the realistic world *dualistically*—a single matter field plus many independent minds—Spinoza's view of the real world was *monistic.* He considered all aspects of the world to be based on a single underlying order. Thus he thought that the interpretation of the physical world in terms of truly independent minds and matter was illusory. That is to say, his was a philosophic view in which the 'subjective' aspects of the world (those that might be identified with independent consciousnesses) and its 'objective' aspects are not really separate. This is in the sense that mind and matter are simply different manifestations of one world, based on the same set of universal principles. Not all of these principles are what we now call 'physical'; some of them relate to what we now call 'metaphysical' underpinnings of reality. Still, the Spinozist view is that both the physical and the metaphysical aspects of the real world belong to the same set of basic truths; that is, they are on the same footing as *primary truths.*[18] This view of a basic oneness of all that there is is not really new with Spinoza.[19] One sees this in much earlier writings, such as the important treatise of the Medieval theologian and philosopher, Moses Maimonides, called *Mishna Tora,* in which he attempts to logically explain the Old Testament.[20] His book starts with the premiss that *all* is one. This view was expressed orally centuries before Maimonides, in the ideas of Jewish mysticism, later written in the *Zohar.*[21] But Spinoza forcefully imposed this view in the Renaissance period in Europe, helping to shape a particular view in science which was to become very important in the twentieth century in Einstein's theory of general relativity.

David Hume (1711–1776)

In the century that followed Galileo, Descartes and Spinoza, David Hume in Britain exerted an important influence on the philosophical approach of the scientific community. His view is sometimes called 'the skeptical approach'. He started his inquiry into precisely what it is that we can know, by asking: are we justified in reasoning from repeated instances of which we have experience to other instances of which we have no experience? He answered that on logical grounds alone the reply must be *no.* But then he continued with the second question: Why, nevertheless, do all reasonable people expect, *and believe,* that instances of which they have no experience will indeed conform with those of which they do have experience? For example, though we have seen the Sun rise every morning of our lives, why do we *believe* that it will rise again in all future mornings?

Hume's answer to the latter question was: because of habit; that is to say, we believe it because the human being is conditioned by repetitions and associations of ideas. According to his logical considerations, then, Hume claimed that all that we can know must be based on past experiences. However, based on psychological considerations, he asserted that the human being does possess in his consciousness a feature called 'habit', that gives him extra knowledge about events that have not yet been experienced. Accepting both conclusions as correct, Hume then arrived at a logical dichotomy. This dichotomy is generally called 'Hume's problem'.[22]

Immanuel Kant (1724–1804)

Kant tried to solve Hume's problem by asserting that some human knowledge does indeed transcend the facts of experience.[23] He called the latter *a priori* knowledge. It comes, in his view, from a coupling of some aspect of the world, called a 'thing-in-itself', with human consciousness. Thus he believed that one aspect of human knowledge is independent of experience. It deals with conventions, analogous to ordinary language, that we impose in our minds and the logical constructs that they entail, such as the rules of a chess game. This type of knowledge is called 'analytic'. Such knowledge is *necessarily true* because conclusions here follow according to invented rules and a set of initial assumptions that are also invented. Another type of knowledge that human beings have is called 'synthetic'. This has to do with the real world, whether or not human beings are thinking about it. This knowledge is not necessarily true; it is contingent on nature. Scientific truth is of this sort, since it depends on the real world. Analytic knowledge entails invented axioms and a set of conclusions that follow from them in accordance with an invented set of rules of logic. The conclusions (of true and false) in this sense are automatically built into the axiomatic basis that we start from and the logic we use. However, in the case of scientific truth, this is only provisional. Future experimentation could easily reveal that the conclusions from some scientific theory are not compatible with the actual facts of nature. In such cases, which have arisen repeatedly throughout the history of science, one must reject (or change in some way) the initial hypotheses that went into the theory, hypotheses that led to particular conclusions about particular phenomena.

Kant's philosophy was then intended to explain the 'habit' that Hume referred to in answering his 'psychological question'. Kant took Hume's term 'habit' to be *a priori* knowledge, not based on logical constructs or rooted in any sort of experiential knowledge. To illustrate this sort of *a priori* knowledge, Kant may have said that the human mind *knows* that space is three-dimensional and that points relate to points in terms of the rules of Euclidean geometry. He felt that this knowledge *comes to us* simply by virtue of our human presence in the world.

Of course, Kant was unaware that there are other possible (non-Euclidean) geometrical systems; some of them were discovered in the nineteenth century, after Kant's time. He was aware, however, that Euclidean geometry is a self-contained set of axioms and theorems and an invented logical system of rules—that it is not based on any observational facts. Thus, there is no a priori reason why 'physical space' is necessarily governed by the rules of Euclidean geometry. However, this geometrical system worked well in the practical and scientific problems of Kant's time. It wasn't until the early part of the twentieth century that physicists were forced to abandon Euclidean geometry as an inadequate logic to express a valid theory of the gravitational force, according to Einstein's theory of general relativity (as we will discuss further in chapter eight).

Einstein acknowledged that both Hume and Kant had an important influence on his work in theoretical physics. Still, the central theme of Einstein's life work is not at all based on Hume's skeptical approach or on Kant's notion of *a priori, transcendent* knowledge. Einstein's theory of general relativity, as a basis for a general theory of matter, is rather based on the philosophic stand of abstract realism and the notion of *objective knowledge.* This sort of knowledge is an abstract set of ideas that, in some way, correspond with fundamental aspects of the real world—a view similar to Plato's cosmology. With this approach, to test the validity of the proposed hypotheses (the 'abstract' ideas about the world) one must use some prescribed logical method (though without claiming that there is only one possible logic). One then deduces particulars from a universal (a fundamental law) that, in turn, is a matter of making predictions of observable facts from the proposed law of nature.

There is still a place in Einstein's philosophy for Kant's notion of transcendent knowledge—but only in the sense that it refers to an inquirer's 'intuitive' knowledge. This is a set of feelings that one scientist or another may have about the way in which certain facts of nature may be explained *in a rational way*. These 'feelings' are by no means to be taken as certain knowledge. They must be tested with a logically consistent set of axioms and procedures in order to arrive at conclusions. But it would be fallacious to conclude that one can *automatically know something*, with certainty, based on transcendent knowledge that we may acquire simply by virtue of one's existence in the world.

From Spinoza's viewpoint, Hume's problem is not a problem. This is because of the Spinozist view that eliminates any basic demarcation between a 'knower' and the 'known'—that is, between the subjective and the objective aspects of human knowledge. When a line is drawn between them, it is only for the practical convenience of describing a particular situation. But in Spinoza's view, any physical system has a *totally objective* representation—because of the relativity of the part of the system we may wish to call by the name 'observed' *(object)* and the part we would then call by the name 'observer' *(subject)*. That is, by interchanging the roles of the subject and the object, the full description of the whole, which is nothing more than 'observer-observed', cannot change: *object* becomes *subject* and *subject* becomes *object,* while the total set of relations that underlie the laws of nature remains the same (invariant).

The *Spinozist* view of the real world is fully objective because it relates to a *closed system* at the outset. Such a view rules out, as illusory, any view that would be anthropocentric, geocentric (or anthing-centric!) It is an approach that is conceptually similar to Galileo's generalization of the Copernican view in astronomy, the view that the universe is Sun-centered rather than Earth-centered. In Galileo's view (in terms of 'Galileo's principle of relativity'), it would be meaningless to assert that there is any special center of the universe—Earth, Sun, Galaxy, Man, or anything else. Galileo's principle of relativity was then generalized in Spinoza's monistic approach to the world, which incorporates all of its aspects, including mankind. This philosophic approach was indeed (probably) the most important precursor to Einstein's principle of relativity—the fundamental axiom that underlies the theory of relativity as a fundamental theory of matter.

Two

Nineteenth-Century Physics: A Conceptual Bridge to Modern Views

Attitudes

After the successes achieved by Galileo and Newton in the seventeenth century, many scientists felt that essentially all of the fundamental principles of physics that underlie the workings of the material universe had been discovered. The notions of atomism, action-at-a-distance, determinism (the claim that all of the material constituents of the universe have predetermined paths), were all ideas that would never be toppled. Indeed, it was supposed during the 200 years that followed the period of Galileo and Newton that thenceforth the obligation of the scientist was only to apply these *true* principles, using the new mathematics that was discovered by Newton (the 'calculus' for representing nonuniform motion), in order to discover more of their implications. However, for all practical purposes, it was taken for granted that there were no new fundamental principles in physics to be discovered.

It was because of this unquestioning attitude towards the underpinnings of fundamental knowledge that the physicist generally broke away from the type of thinking that is called 'philosophical', turning towards mathematics for a guide to express his experimental results. Even the name, 'physicist' (previously 'natural philosopher') was changed to 'mathematician'.

I believe that this was a very unfortunate shift of allegiances and that it seriously attenuated real progress in the scientists' quest for an understanding of the material universe. For it was an attitude that abandoned the *critical approach* of the philosopher for the more formal, elegant expression of the mathematician in representing the laws of nature. While the new mathematical language was certainly an important development for the purpose of expressing and exploiting the ideas of physics, the ideas themselves could only move forward extremely slowly without the critical approach of philosophy. This abandonment of the critical attitude in the latter part of the nineteenth century then led to increasing dogmatism about the *underlying ideas,* even in the face of mounting experimental and theoretical refutations of these ideas.

These difficulties, which appeared at first around the end of the nineteenth century, seemed like small gray clouds in an otherwise unblemished clear blue sky. But before a few more decades had passed they grew very large and eventually overturned the existing ideas with the birth of two major upheavals in physics—the quantum and relativity theories. Nevertheless, the dogmatism of nineteenth-century physics, based on the Newtonian approach, was replaced with a different dogmatism. Some of the new ideas came to be just as dogmatically held as were those of the classical ideas of nineteenth-century physics. I believe that the dogmatic *approach* remained in physics in our own time primarily because physics stayed apart from philosophy and its critical attitude.

Returning now from the attitudes of science to the developments of nineteenth century physics, rumblings were heard in that period in regard to both the theoretical and experimental bases of Newtonian physics. Controversy between atomism and continuity, and between the positivistic and realistic approaches to knowledge loomed large. Each of these controversies played the important role of precursor for the developments in the two main changes in nineteenth-century physics—the quantum theory, to underlie atomic phenomena, and the theory of relativity. The quantum theory developed into an approach to elementary matter based essentially on atomism and logical positivism, while the theory of relativity developed into a stand based essentially on the continuous field concept and an epistemological view of realism, the idea that there is a real world, characterized by fundamental principles, and laws, *irrespective of observers*.

Nevertheless, the contemporary quantum theory abandons Newton's 'determinism' while the view of relativity theory is fully deterministic in the sense that all of the fundamental variables of a material system are taken to be predetermined at the outset. In addition, the theory of relativity, as a fundamental theory of matter, is a continuum view of matter that grew out of Faraday's field concept in nineteenth-century physics. We will now review some of the explicit developments that led to the support of atomism, on the one hand and continuity on the other, as they emerged in nineteenth-century physics.

Nineteenth-Century Physics of Matter

In the early years of the nineteenth century it was believed by most scientists that a general explanation for the constitution of matter must be based on the ideas of the gas laws—Boyle's law, Charles' law and their combination in the form of the 'ideal gas law'. The latter was successful in revealing the way in which the volume of a gas, V, depends on its temperature τ and its pressure P. Defining the temperature scale as 'Celsius', water would freeze at $0°$ C and it would boil at $100°$ C. It was discovered in the nineteenth century that the temperature of matter has a lower limit, at $273°$ C below the freezing temperature of water, where the pressure and volume of the gas vanish. The ideal gas law then took the form:

$$PV = R(\tau + 273) \quad (\to 0 \text{ as } \tau \to -273°C)$$

where R is the 'universal gas constant'. Its value is independent of the particular gas that this law describes.

It is conventional to use a more practical temperature scale, in degrees Absolute by shifting the Celsius scale by $273°$. That is, with $T(°A) = (\tau + 273)(°C)$, we may write the gas law in the simple form:

$$PV = RT$$

The parameters that appear in this 'ideal gas law' then have the following meaning: V is the spatial volume that would be occupied by the gas in a container, bounded by walls that separate this quantity of gas from the rest of the outside world. The parameter P is the pressure exerted by the gas on its containing walls. Dynamically, 'pressure' is the force exerted on the walls of the container, perpendicularly, per unit area. The parameter T, at this stage of the description, is then the reading of a thermometer, in °A.

It was clear that this form of the ideal gas law was strictly phenomenological, that is to say, descriptive of the relations between volume, pressure and temperature, *though not explanatory*. The question that arises is: What are the *fundamental* meanings of the parameters, V, P and T? We have already said that the pressure is the force per unit area that acts perpendicularly on the walls of the container that separates the gas from the rest of the outside world. But what is the nature of this force? Secondly, *why* does the temperature of the gas have an absolute minimum (though not an absolute maximum)? And why is it that the pressure and volume of the gas reduce to zero at the minimum temperature, $T = 0°A$? To begin to answer these questions, let us start with an important discovery on the atomistic nature of the gaseous state of matter.

John Dalton (1766–1844) and Atomism

In the early part of the nineteenth century, the eminent chemist John Dalton observed that if a gaseous mixture should be depleted of one of its constituent ingredients, such as when all of the nitrogen is removed from some volume of air (recall that air is mainly a mixture of nitrogen and oxygen), a sudden change in the pressure of the remaining gas is produced. This result led Dalton to deduce the 'law of partial pressures': the assertion that the actual pressure of a gas is a sum of partial pressures of its constituent ingredients.

The pressure of the gas is caused by the large number of atomic (or molecular) constituents striking the walls and bouncing backward from the collisions, repeatedly. According to Dalton's model, then, the *sudden* change in pressure of the gas, when a particular component of the gas mixture is removed, is because of the disappearance of particular atomic constituents, *each with the same discrete inertial mass*. Thus, Dalton concluded that matter is generally composed of discrete atoms, each type with its characteristic mass, and thereby gave support to the atomic nature of matter.

Conservation of Energy

The concept of energy was developed in the nineteenth century, along with the law of its conservation. It is interesting to note that some of the scientists who were concerned with this problem, such as Julius Mayer, were motivated by the following question in physiology: what is the source of the heat of a living body? This led to the additional question: what is heat, in fundamental terms? Some said that heat, *per se,* is an atomistic phenomenon, whereby it is carried by 'particles of heat', called 'caloric particles'. But the pioneers in this field, Mayer, H. Helmholtz and J. Joule, discovered that in fundamental terms, heat is a particular form of 'energy' which, by definition, is a measure of the work that some quantity of matter or radiation is capable of doing.[24]

'Energy' was found to manifest itself in several different ways, in addition to the form of heat. There is energy of motion, called 'kinetic energy', 'potential energy' due to the spatial position of a quantity of matter (for example, the gravitational potential energy of a block of wood, elevated above the Earth's surface, or the electrical potential energy of a charged body due to its spatial position relative to other charged bodies), 'chemical energy', and so forth. What was discovered was that the *total energy* of a system must be conserved in time; that is, its total capability of doing work is a constant. Thus, even though some of the electrical energy that pushes a current through a coil in a beaker of water is converted into heating up the water, the sum of the electrical and heat energies stays constant. Another example is the conversion of some of the potential energy of a rock at the top of a cliff into kinetic energy as the rock falls to the lower level. At any given time during the fall, the sum of the potential energy of the rock and its kinetic energy is a constant, the potential energy continually diminishing and the kinetic energy increasing.

The law of conservation of energy was *derived from* the cause-effect laws of motion of Newton, for the motions of the constituent atoms of a material system, though not until these conservations relations were discovered in terms of empirical facts. The conservation of energy law was then tied to the laws of thermodynamics, particularly the first law.

The Laws of Thermodynamics

The combination of the law of conservation of energy and the identification of heat with a form of energy led automatically to *the first law of thermodynamics:*

> Heat injected into a material system must be used up in two ways: 1) in increasing the internal energy of the system and 2) in causing the system to do work against its environment, such as a gas pushing the walls of a container (if they are moveable) so as to increase the volume of the container.

The second law of thermodynamics concerns a *change* of a material system of more order, when it is not in an equilibrium state, to one of less order, when it evolves toward the equilibrium state, reaching the maximum disorder at equilibrium. The law says explicitly:

> If a system is in a state of non-equilibrium, then if left on its own it will proceed *irreversibly* toward the state of equilibrium. Such change corresponds to an evolution from more order to one of less order *in time,* until the maximum disorder would be reached, at the state of equilibrium.

To exemplify the latter law, consider that all of the air molecules in a room are squeezed into one of its corners. This would correspond to a more ordered state of the system since one would be able to say with more certainty where any of the air molecules are located, compared with the situation when they are all over the room, uniformly. The latter case, which corresponds to equilibrium, entails the idea that any of the air molecules could be anywhere in the entire room, thus there is more uncertainty (disorder) involved in the specification of the locations of any one of the molecules, thus there is more disorder. The second law of thermodynamics then asserts that when the molecules of the room are squeezed into a corner of the room, and left on their own, they must diffuse into the room so as to occupy all of its volume, uniformly *and permanently* if left alone. This process would then be one of going from more order (at nonequilibrium) to one of less order, until maximum disorder is achieved, and maintained, at equilibrium.

The quantitative measure of disorder, that in turn relates to the thermodynamic functions, temperature and heat, is called 'entropy'. The quantitative expression of the second law of thermodynamics then asserts that at nonequilibrium the entropy of a system has less than its maximum value, but that it would then evolve *in time* to its maximum value, achieved at equilibrium, where it would remain for all future time if left on its own.[25]

Ludwig Boltzmann (1844–1906) and his Statistical Approach to Matter

In the latter decades of the nineteenth century, Boltzmann was concerned with *explaining* the physical features of gases that were known only in *descriptive* terms at that time.[26] This was a step beyond describing the gaseous state (as with the ideal gas law) and toward understanding this law in terms of cause-effect relations. Boltzmann's starting point was the atomic model of matter. Assuming that Newton's force laws were the true cause-effect relations that govern the motions of these atoms, Boltzmann then concluded that all of the properties of a macroscopic quantity of a gas should follow from the physical properties of this very large number of atoms, moving about under the influences of their mutual Newtonian forces, *acting at a distance* in causing each other to have the particular trajectories that they have in space, even though the individual atoms of the macroscopic quantity of the gas is invisible to the human eye (or invented instruments). Boltzmann believed that, for example, the ideal gas law must be based on such an underlying set of cause-effect relations.

In regard to the number of atoms in a macroscopic quantity of a material system, it was discovered also in the nineteenth century, in the chemical researches of Amadeo Avogadro, that a macroquantity of matter, called a 'mole', always contains the same number of molecules or atoms. The 'mole' of substance is defined as that weight of substance in grams that is equal to the numerical *atomic weight* of that substance. That is to say, 2 grams of H_2, 32 grams of O_2, 14 grams of N, etc. are each a mole of that substance and each contains the same number of constituent molecules (or atoms). That number was found to be the order of 6×10^{23}. Thus, 32 grams of diatomic oxygen, O_2, 2 grams of diatomic hydrogen, H_2, 14 grams of atomic nitrogen, N, each contain 6×10^{23} constituent molecules or atoms of the respective substances.

In analyzing the physical properties of macroscopic quantities of such gases, Boltzmann was then confronted with the horrendous task of writing down 6×10^{23} coupled differential equations for each constituent molecule of a mole of substance—Newton's equations of motion for each of them, specifying six boundary conditions for each of these molecules (their positions and velocities at some initial time) and then determining their solutions. Of course, no human being (nor even computing machines that might be compounded of all existing computing machines on Earth) could ever carry out this impossible mathematical task, even though it can easily be proven that there exist solutions of these equations. Boltzmann then decided to take a different route toward the description of a macro-quantity of matter. His research program was to find the equation whose solutions would be the proper *weighting function* that would allow one to calculate the *average properties* of a very large ensemble of atoms or molecules, such as their average speeds, average potential or kinetic energies and so forth, without the need to know their individual speeds, potential or kinetic energies, etc. The equation was called 'Boltzmann's equation' and its solutions were called the 'distribution function'. This is like saying that if a six-sided die is made of a homogeneous material, one can make the prediction that, *on the average,* out of a very large number of throws, it will land with the number 'two', one-sixth of the time. This conclusion did not depend on the dynamics of the thrown die. It only depended on a probability theory in which it follows that the probability of landing 'two' is 1/6, if all six sides of the die are equally weighted. Still, there is a dynamics that underlies the motion of the thrown die, whose equation of motion predicts precisely which side the die will land on given the precise boundary conditions of initial position and speed of the thrown die. But because it is extremely difficult to carry out this particular mathematics and find out the precise boundary conditions, one resorts to asking probability questions instead of questions regarding certainty. If this were not the case, it would be possible to win a great deal of money at the dice table of a gambling casino because the outcome of the thrown dice would no longer be a gamble; it would be a certainty.

Thus Boltzmann discovered from purely statistical arguments that the product of the pressure of an ideal gas and its volume was as follows:

$$PV = (2N/3)\langle K\rangle$$

for a 'mole' of a gas, where N is 'Avogadro's number', equal to 6×10^{23} atoms (or molecules) per mole, and $\langle K\rangle$ is the *average* kinetic energy of the atoms (or molecules) of the gas.

If we compare this result with the empirically determined result for the ideal gas, $PV = RT$, we may identify the average kinetic energy/atom of the gas with its absolute temperature:

$$T = (2N/3R)\langle K\rangle \text{ °A} = (2/3k)\langle K\rangle \text{ °A}$$

where $k = R/N$ is called 'Boltzmann's constant', numerically of order 1.4×10^{-23} *J/°A*. Thus Boltzmann made the remarkable discovery that the temperature of a gas, in °*A,* is nothing other than a measure of the average kinetic energy of the constituent atoms or molecules of that gas. That is, if the internal energy of the gas should be slowly decreased, thereby slowing down its atoms' motions, on the average the temperature of the gas would correspondingly decrease. The absolute minimum temperature of the gas, $T = 0$°*A,* then corresponds to the situation in which all of the constituent atoms have *stopped in their tracks*. They would then be sitting in a small pile at the bottom of the container. In this limit, the volume occupied by the gas would be essentially zero and its pressure would be zero, since there wouldn't be any atoms hitting the walls of the container at this minimum temperature, *0°A*.

In our discussion of the *first law of thermodynamics*, it was indicated that at least a part of the heat energy injected into a material system will be used up in increasing its internal energy. What Boltzmann discovered, then, was that this corresponded to stepping up the internal (kinetic) energies of the constituent atoms *(on the average)*, thereby causing the temperature to rise.

Later on it was discovered that the relation between average kinetic energy of the atoms of the gas and the temperature (in °A), $\langle K \rangle = (3/2)kT$, is only a special case of a *general law of equipartition of energy*. This is the assertion that the average energy (of any sort), $\langle E \rangle$, of the constituent atoms or molecules of a large ensemble, per degree of freedom of the atoms, is just equal to $kT/2$. The factor of 3 in the preceding relation for the average kinetic energy was due to the three degrees of freedom in the motion of the atoms in 3-dimensional space. But if these would be diatomic molecules, there would be two additional degrees of freedom—those associated with the vibrations of the atoms relative to each other in the molecule (one vibration along their line of centers and the other perpendicular to it). In this case, as would apply to O_2 or H_2, the average energy per molecule of the gas, at temperature T, would be $\langle E \rangle = (5/2)kT$ *Joule*.

The latter predictions were in very good agreement with the data in nineteenth-century physics. Deviations were detected later on into the twentieth century, at very low temperatures. These were then explained with the new quantum mechanics that was to emerge, and the different sort of statistics that went with it. But the agreement between theory and experiment in the nineteenth century did give strong support to Boltzmann's atomic theory of matter, as well as Dalton's original conclusions about the fundamentally atomistic nature of matter.

In regard to the *second law of thermodynamics*, Boltzmann represented the 'entropy' of matter in terms of a probability function, relating to the probable distribution of states of a many-body system of matter. Denoting the entropy by *S*, he found that it was linearly proportional to the natural logarithm of the function *W*,

$$S = k \ln W$$

where *W* is the probability of a particular distribution of an ensemble of atoms or molecules, among all of their possible states of motion. When $\overline{S}$ is maximum entropy, i.e., disorder, $\overline{W}$ then corresponds to the most probable distribution of available states for the gas. With this expression for the entropy of a many-body system in terms of a distribution of states of probability it was then proven that when a system (a material ensemble) is in a non-equilibrium state (a state of less than the maximum disorder), its entropy must increase irreversibly *in time,* toward the state of maximum disorder, achieved at the equilibrium state. At the equilibrium, if the system is left on its own, it should remain in the state of maximum disorder forever. Thus, Boltzmann succeeded in predicting the consequences of the second law of thermodynamics from a particular sort of probability calculus.[27]

At this stage of the discussion it is important to note once again that Boltzmann's expression of the second law of thermodynamics did not address the problem about the underlying cause-effect relations (dynamics) for the constituent atoms of the gas. It only concerned the empirical fact that complex systems proceed irreversibly toward equilibrium, describing this evolution in terms of proceeding from order to disorder. But such disorder, in Boltzmann's theory, had to do with a lack of knowledge on the part of the inquirer about the complex system. For it was assumed with this theory of the nineteenth century that there *does* exist a precisely *predetermined* set of states of the entire ensemble, as predicted by Newtonian physics. With this view, the second law of thermodynamics concerns the lack of knowledge of the inquirer about all of the details of the physically evolving system. For example, if a clear blue ink drop is inserted into a colorless liquid, there would be, initially, maximum order, for it would be possible to specify fairly precisely where all of the blue ink molecules are located. But in time, the ink molecules diffuse into the liquid until the entire (formerly colorless) liquid is a pale blue, uniformly. At this stage (of equilibrium), one would be maximally uncertain about the whereabouts of any particular ink molecule; it could be anywhere in the entire liquid. Thus there is maximum entropy here (maximum disorder). But the forces between the ink and colorless liquid molecules were the *cause* of the diffusion of the ink drop into the liquid. The paths of each of the ink molecules were predetermined, even though the inquirer did not have a precise knowledge about these paths. Thus, the probability statements that accompany the second law of thermodynamics are tied to our knowledge about a system; they are not tied to the objective evolution of the system.

It should also be noted that the meaning of 'time' in the context of the irreversible evolution of a complex system toward equilibrium is different than the meaning of 'time' in the description of the trajectory of a single atom of matter. Both 'parametric change measures' are called by the name, 'time', only because they may be correlated *in special cases*. But it is indeed important to keep in mind the fact that these are quite different concepts of 'time' (as different as apples and trees—even though one may in some way correlate apples and trees, they are quite different entities!). A similar confusion arises in regard to the time concept as it is defined in the theory of relativity; this will be discussed in detail in chapters eight and nine.

Boltzmann Versus Mach

During the period of Boltzmann's triumphs for the atomistic model of matter, one of his colleagues and intellectual adversaries, Ernst Mach (1838–1916), proposed quite different views. In contrast with Boltzmann's philosophic stand of realism (though it was 'naive' realism rather than 'abstract' realism), Mach's philosophic stand was based on positivism. These opposing epistemological points of view, which we have discussed previously, are the same differences that have persisted among scholars in philosophy and science since the ancient periods in the Western and Oriental civilizations.

Boltzmann believed that matter must be atomistic because on the macroscopic scale our minds perceive the effects of individuated quantities of matter. Because of this perceptual knowledge, he believed that the human mind must also have some transcendent knowledge about matter on the microscopic scale, where it has no perceptual knowledge. In addition it was Boltzmann's understanding that we should gain such transcendent knowledge because of the way that the world is. That is, the way nature is in the domains where our sensing apparatuses don't respond must necessarily be an image of the way nature is where our sensing apparatuses do respond. This view is called 'naive realism'.

In the early part of the twentieth century, Bertrand Russell claimed that naive realism was philosophically indefensible. His argument was as follows: "The observer, when he seems to himself to be observing a stone, is really, if physics (physiology) is to be believed, observing the effects of the stone upon himself. Thus, naive realism, if true, is false; therefore it is false".[28]

Boltzmann also professed that *all* hypotheses are, in principle, refutable. This should then include his own hypothesis about the atomistic nature of matter. It was then his 'philosophy of science' that to investigate the laws of the physical world one must continually *test* the existing paradigms. Such a philosophy of science was carried forward in our own time by Karl Popper, who laid more stress on the refutation than the verification of scientific theories, in his approach of 'conjectures and refutations'.[29]

In contrast with Boltzmann's naive realism, Mach asserted that all one could claim to be real are the reactions of the human being's senses to the outward manifestations of matter. It was then his view that the job of the scientist is strictly to classify the data of scientific experimentation as economically as possible, in order to ascertain the regularities in nature. But he did not believe that it is meaningful to extend these regularities to ideas that are claimed to *underlie* them. It was Mach's view, because the human being does not directly perceive the atoms of Boltzmann's theory, that such 'atoms' were not more than a mathematical artifice—a 'tool' to facilitate our description of material phenomena.

Nevertheless, along with Boltzmann, Mach did believe in an *anti-dogmatic* approach in the philosophy of science. In his treatise, *The Science of Mechanics,* Mach expressed this view as follows:[30]

> The history of science teaches that the subjective, scientific philosophies of individuals are constantly being corrected and obscured, and in the philosophy or constructive image of the universe which humanity gradually adopts, only the very strongest features of the thoughts of the greatest men are, after some lapse of time, recognizable.

To sum up their conflict, Boltzmann and Mach had fundamental disagreements on epistemology, though they agreed in principle on the method of approach to scientific facts. Boltzmann was an atomist and philosophically a naive realist. Mach did not accept the atomic model as fundamental and he took a positivistic philosophic stand. Nevertheless, both scholars were anti-dogmatic in their philosophies of science.

In our own day, the most widely accepted theory of matter—the quantum theory of measurement—is based on the notion of atomism (Boltzmann), while its epistemology is positivistic (Mach).[31] However, unlike the common anti-dogmatic view of Mach and Boltzmann, quantum mechanics takes a dogmatic stand. It makes a bridge between Boltzmann's atomism and Mach's positivism by rejecting the determinism of Newtonian physics, which is embedded in Boltzmann's physics. That is, with this modern theory of physics applied to elementary matter, one no longer assumes that the properties of the elementary bits of matter, such as their trajectories, are predetermined. One rather assumes that the physical features of elementary matter must be in accordance with the way in which a measurement is carried out *with a macroscopic apparatus, when it is carried out.*

The brand of positivism adopted in the quantum mechanical view is somewhat watered down from that of Mach. The altered approach, called 'logical positivism', was worked out in the early decades of the twentieth century, by a group of philosophers in Austria (Boltzmann's and Mach's home ground) known as the 'Vienna Circle', in Germany (the 'Berlin Circle'), and in other centers in Europe and the United States. An essential theme of logical positivism is its underlying 'principle of verifiability',[32] the assertion that *all* formal statements within the language of science must be verifiable, though not necessarily by the human senses. Any other statement, in the context of science, is claimed to be 'meaningless'. For example, to say that the potential field of force of an electrical charge is continuously distributed in all of space would be called 'unscientific', since it is not a statement about the actual force exerted on a test body. The latter is, rather, *derived from* the former (more abstract) statement. From the standpoint of logical positivism, then, it is required that all 'meaningful' statements be formulated in logically consistent language structures, and relate directly to the data of scientific experimentation.

The present-day quantum theory of measurement rejects the anti-dogmatic approach of Boltzmann and Mach because of its claim that *the data are the theory and the data cannot be false.* Thus the view of quantum mechanics is often claimed to be a final truth about the fundamental nature of elementary matter; therefore it claims to be irrefutable. According to Boltzmann's (and Popper's) philosophy of science, this would imply that quantum mechanics is not a scientific theory! Further discussion of the details of the quantum mechanics approach will be given in chapters six, seven, and ten.

From the philosophical view, Newton's determinism is not abandoned in Boltzmann's theory of matter—probabilities were useful to him for determining averages of very large ensembles of atoms that make up a macroscopic quantity. However, as we will see in later chapters, the probability calculus of quantum mechanics was interpreted as fundamental. That is, the laws of nature fundamentally became laws of probability, assumed to underlie the very nature of elementary matter.

In this sense, what we came to with the revolution of quantum mechanics was the new and fundamental role played by probabilities in the laws of matter. Rather than the predetermined space and time co-ordinates and velocities of the individual atoms of matter that comprise a physical material system, the basic variables of these atoms become *probability functions* of the space and time co-ordinates. Such a theory is called 'nondeterministic', because trajectories are not 'predetermined'.

Faraday (1791–1867) and the Field Concept

As significant as the atomistic models of matter put forward by Dalton and Boltzmann was the continuous field concept, put forward in that same period. Michael Faraday's main idea was to say that *instead of* considering that matter is a collection of 'things', each with their derivative physical properties, one of which is their field of continuously distributed potential influence on a 'test body', it is rather the continuous field of potential influence that is fundamental to the nature of matter, with the 'thing' attributes being derivative to the field. Thus, in Faraday's view, the feature of matter that makes it appear as though it were localized is in fact a derivative property of the *underlying* field that expresses the essence of this matter.[33]

To exemplify Faraday's notion, consider once again Newton's formulation of the law of universal gravitation from Faraday's field perspective. Newton discovered that his second law of motion, when applied to the gravitational force in particular,

$$F = ma = GmM/r^2 \quad (\text{where } a = d^2r/dt^2)$$

led to all of the observed features of planetary motion as we have discussed in the preceding chapter, as well as other of the known gravitational and terrestrial effects, previously thought to be independent of the causes of the planets' motions. While Faraday accepted the *empirical* success of Newton's equation, he did not believe it was imperative to accept Newton's interpretation of this equation in terms of material things, localized at discrete places and acting on each other at a distance, spontaneously.

Instead of Newton's action-at-a-distance idea, Faraday proposed that his equation could equally be interpreted in terms of a continuous field of potential action

$$P(R) = GM/R^2$$

This field is defined everywhere in space, except at $R = 0$. It represents a 'potentiality'—for the action of one massive body, from its center where it may appear to be a discrete body, on any test body anywhere, say at the spatial location R. Thus, the *actuality* manifested by the test body with mass m is determined by the coupling of m to $P(R)$ at the place R where the test body is defined to be. That is, if the test body is at the place $R = r$, then

$$F = ma = mP(R = r) = m(GM/r^2)$$

thus giving the identical law that Newton derived, though from a continuum point of view rather than the view of localized atoms of matter, where the continuum field, $P(R)$, is the elementary starting point for a basic representation of the matter. $P(R)$ is a 'field' because it is continuously distributed throughout all of space, except for the point $R = 0$. To include this point in Faraday's field would be illogical since it would describe a field P acting on its own source, whereas the field in Faraday's definition is to represent the effect of the source of this field on *other* matter.

The concepts of 'self-force' and 'self-energy', which do appear in contemporary elementary particle physics, are automatically rejected by Faraday's field theory in terms of *potentialities*. Recall that also in Newton's laws of motion and in the expression of 'Galileo's principle of inertia', the term 'force' always refers to the action of *external* bodies in causing a non-uniform motion of the body under consideration.

According to Faraday's field theory of matter, what the atomist would call an n-body system must be fundamentally described as a superposition of n continuously mapped fields of potential force. The resultant field of potential force would then fill all of space (generally non-uniformly). This field would just be there, to act on a test body should it be somewhere in space to be acted upon.

A logical difficulty with this approach was that the test body itself was still to be treated as a 'particle' of matter assumed not to exert any influence on the source of the field of force that acts on it. This seemed to be contrary to the spirit of Newton's third law of motion, which asserts that for every action in the world, in terms of one force that acts on a body, there is an equal and opposite action of the body acted upon, in affecting the source of the force that acts on it. The asymmetry in the fundamental description of the field of force and test body, in Faraday's field theory, was not to be resolved until the revolutionary development that came with the theory of relativity in the twentieth century.[34]

When Einstein's theory was developed to the stage of general relativity, there was a basic conceptual change from Faraday. For Faraday the material world was a large collection of individual entities, though the variables used to describe them were continuous fields of force rather than localized variables of the atoms of matter. His system was fundamentally *open.* But with the theory of relativity, in contrast, we come to the view of a material system as fundamentally *closed*—that is, a single entity without parts, even though it does manifest itself as seemingly individual modes of behavior, appearing to us as though there are indeed separate objects. These are 'modes' of the closed system, in this view, analogous to the ripples *of* (not in) a pond. Though the ripples of the pond are certainly individual entities, they are not really atomistic in the sense of being separable from the pond. Such is the view of a *closed system* that came with Einstein's theory of general relativity, as we will discuss in more detail in chapter eight. However, it can be seen at this stage how Faraday's field concept was a very important precursor for Einstein's field concept in his general theory of matter. It passed from the concept of an open system to that of a closed system for the underlying understanding of the universe in any of its domains—from elementary particle physics to cosmology.

Faraday first applied his field theory to explain electricity and magnetism. The electric field of potential force between positively and negatively charged plates of an electrolyte was, to him, the *essence* of those plates. He saw the manifestation of these plates as positive and negative charge that attract or repel the ions in the fluid (the 'test bodies') with the positive ions moving toward the negative plate and the negative ions moving toward the positive plate, each moving along a continuous *electric line of force.*

Since, in Faraday's view, the electric line of force is the primitive representation of the charged matter, it followed that for every positive electrical charge in nature there must also be a negative one—simply because a line necessarily has two ends—defined here as electrical charges with opposite polarity.

When Faraday applied his field theory to magnetism, he found that the continuous mapping of the magnetic lines of force (say, that would align a compass needle) is in terms of lines that *bend,* so as to curve back toward their point of emanation. The reason for this is that the magnetic field was found to be fundamentally *dipolar;* for example, a magnetic line of force that leaves a North Pole will bend around to eventually return to the South Pole associated with it. The electric lines, in contrast, diverge outward from *monopoles* of a particular polarity, only to meet a charge of opposite polarity somewhere out of the domain of the first charge. Thus, it was discovered that the smallest amount of magnetic material maintains this inseparability of sets of two poles. If a magnetic material should be cut into smaller and smaller pieces, each new cutting would still leave a North Pole and a South Pole just as there is a type of worm that grows a new head (or tail) whenever it is cut in half, so as to yield two new worms, each with a head and a tail.

Faraday next discovered that one can generate an electrical current (a moving electric field of force) in a metal wire if this wire should be moved across magnetic lines of force in the vicinity of a magnet. He then concluded that an electric field of force, that is, the *cause* of the motion of an electrically charged particle of matter, is *equivalent to* a magnetic field of force in motion. It also followed that, reciprocally, a magnetic field of force is *equivalent to* an electric field of force in motion. Now since motion *per se* is purely subjective in the description of interacting bodies, Faraday concluded that the electric and the magnetic fields of force are simply two different manifestations of a single, *unified* field of potential *electromagnetic* force. Thus, Faraday ushered in the notion of a unified field theory, an approach to the fundamental nature of matter that did not re-emerge until the twentieth century, when Einstein attempted to further unify the electromagnetic field of force with the field of the gravitational force, in his theory of general relativity.

From the point of view of the philosophy of science, particularly in regard to epistemology (the theory of knowledge), it should be noted that the electromagnetic field in itself is not a directly observable entity. It is rather a 'potentiality', to be actualized only when a test body may be somewhere in space to feel its influence. Faraday's field theory is then based on the philosophic idea of 'abstract realism' whereby one *deduces* a part of reality of the world from a comparison between particulars that follow from a universal (the underlying field relations) and the observable facts that may be discovered in experimentation. In his theoretical approach the *underlying* electromagnetic field of potential force is an unobservable feature of matter—it is 'metaphysical'—though it is essential for an explanation of the true nature of electrically charged matter. This approach opposes 'logical positivism', which would claim that no such 'metaphysical' quality can be meaningful in science.

According to Faraday's philosophy, then, the real, underlying essence of electrically charged matter is a continuously distributed electromagnetic field of potential force. This approach to realism ('abstract realism') is in conflict with Boltzmann's 'naive realism' in his explanation of matter in terms of an atomistic model. According to Faraday, abstract realism must incorporate *all* possible manifestations of matter in influencing other matter. It is 'abstract' in the sense that what it is that is real does lead, by logical implication *(but not directly through the senses)* to the observable facts of nature. According to this view, as indicated earlier, one starts with a *universal,* then derives particulars from it that are to be compared with the observable facts of nature. One must always keep in mind, with this approach, that some particulars that follow from such universals are logical ingredients, but not observables, though they are *logically necessary* within the structure of the theory.

Some philosophers and scientists argue that this 'alleged' underlying reality, claimed by the realists and expressed in terms of universals, is no more than a way of categorizing the regularities in the data of scientific experimentation, and nothing more than this should be read into it. Such would be the response of a logical positivist. However, it may be argued that this claim is untrue because of the logically necessary relations of the theory that are not in accordance with the principle of verifiability—the particulars of the theory that are not directly observable phenomena. These logical relations of the universal theory are somewhat analogous to the relation of the totality of a woven blanket to all of its threads. Not all of these threads may be observable. Yet, should one of them (observable or not) become defective, the entire blanket would fall apart.

James Clerk Maxwell (1831–1879) and the Laws of Electromagnetism

It was Maxwell's genius that put Faraday's ideas of the continuous field concept into an explicit mathematical form as partial differential equations. But Maxwell did not interpret these equations and their solutions as Faraday did. He saw the continuous fields as derivative features of bits of matter, rather than vice versa.

In addition to predicting all of the electromagnetic phenomena that were studied by Faraday and other important early scientific discoverers of electromagnetic phenomena (Ampère, Oersted, Biot and Savart, Henry, Franklin, . . .), Maxwell discovered further that his field equations for the laws of these phenomena also yielded predictions for all of the optical phenomena known in his day. Thus, the appearance of Maxwell's field equations in the nineteenth century was a giant step forward toward the understanding of electromagnetic and optical phenomena, in terms of a fully unified field theory.[35]

A particular class of solutions of Maxwell's electromagnetic field equations, called 'the radiation solutions', revealed that light is nothing more than a particular electromagnetic phenomenon. It is in the form of a transverse vibration of the electric and magnetic fields of force, which oscillate together in a direction perpendicular to their direction of propagation, and each (electric and magnetic field) perpendicular to each other.

The electric and magnetic vibrations in this propagating 'electromagnetic field of potential force' are each oscillating at the same frequency. Each such electromagnetic wave solution of Maxwell's equation, in turn, corresponds to a particular frequency component.

The spectrum of solutions for the phenomenon of 'visible light' covers a range from the frequency that identifies with deep red (low end of the frequency range) to orange, green, blue, indigo, and then to deep violet (at the high end of the frequency range). The frequencies just lower than 'red' are called 'infra-red'—associated with a heating effect of the sun's radiation. The frequencies just higher than violet are called 'ultra-violet'—they are associated with many physical phenomena, and have effects on biological entities, such as viruses, reviving them from a dormant state. With the relation between frequency, wavelength and the speed of light, $c = \nu$ (frequency) λ (wavelength), and the magnitide of the speed of light, of order $c = 3 \times 10^8\ m/sec$, the frequencies of visible light are of order 10^{15} Hertz, corresponding to wavelengths of order $10^{-6}\ m$. Other radiation solutions of Maxwell's equations, that correspond to wavelengths of order of kilometers, are associated with the phenomenon of radio propagation, the radiation solutions associated with wavelengths of order of a few cm are identified with microwaves (FM radio and TV, as well as 'microwave ovens'), wavelengths of order $10^{-10}\ m$ are X-rays, and wavelengths of order 10^{-15} m are gamma rays. All of these phenomena are special cases following from particular radiation solutions of Maxwell's electromagnetic field equations.

The transverse characteristic of electromagnetic radiation, and light in particular, was found several decades before Maxwell's discoveries, by Augustin Jean Fresnel (1788–1827).[36] The propagation of radio waves was discovered by Heinrich Rudolph Hertz (1857–1894), after Maxwell's discoveries, and the first radio set was invented by Gugliemo Marconi (1874–1937). All of these achievements further substantiated the notion, first propounded by Robert Hooke and Christian Huygens in the seventeeth century, that light is a continuous wave phenomenon. The earlier view was that the light wave is analogous to the sound wave, a longitudinal vibration rather than a transverse vibration. Still, it is significant that since the time of Newton, 200 years before Maxwell's discovery of the relation between light and electromagnetism, the one optical phenomenon that Newton did not really explain satisfactorily, in view of the experimental facts of his day, was optical diffraction. But the decision about the basic nature of light, as to whether it is corpuscular or a wave form, remained inconclusive until the breakthrough achieved by Maxwell. It must have been a pleasant feeling for the physicists of the late nineteenth century to know that, at last, the dispute about light had been settled (though the settlement was not to last more than a few decades).

In spite of the successes of Maxwell's field equations in giving a satisfactory explanation for electromagnetic phenomena in terms of continuous fields of force, Maxwell himself remained an atomist in regard to light propagation. He believed that what was doing the vibrating in the case of light propagation were the atoms of ether that were conducting this light from its source to some absorber. Of course, he knew that light apparently propagates from the distant stars to us through a vacuum. But he believed that all of space is filled by a lattice of ether atoms that indeed conduct the electromagnetic radiation, just as air molecules vibrate as they conduct the propagation of sound waves. Nevertheless, Maxwell was not aware of any experimental evidence for the existence of any of the other physical properties of the ether, such as its viscosity, turbulence, heat conductivity and so forth. Maxwell was disappointed until the last years of his life because there wasn't a single piece of *direct* evidence to support the ether model of light and other electromagnetic radiation propagation. But in conflict with the ether model was Faraday's idea that the electromagnetic field was only the potential throughout space for charged matter to act on other charged matter. This idea did not require the existence of the ether. Faraday's idea of the field indeed came closer to Einstein's view. It is interesting to note that Einstein was born in the same year that Maxwell died.

The Michelson-Morley Experiment

Toward the end of the nineteenth century, Albert Abraham Michelson (1852–1931) and Edward Williams Morley (1838–1923) carried out an experimental investigation that should have provided a precise demonstration of the existence of the ether, if it had existed.[37] The experiment utilized the Michelson-Morley interferometer. The idea was as follows: one sends two beams of monochromatic light (i.e. light with a fixed frequency) in different directions relative to the Earth's surface. If they travel along equal pathlengths, initially in phase with each other, returning to the starting point with mirrors from perpendicular locations in space, they should then be out of phase. The effect would be to observe interference fringes in the recombined monochromatic waves. Assuming that the Earth moves at the speed v relative to the ether and the sun (taken as the reference frame), the speed of light was then expected to be $c - v$ when the light moves away from its source and $c + v$ when moving toward its source. This is similar to the net speed of a boat when moving with or against the stream. Similarly, when the light propagates perpendicular to the Earth's surface, its speed both toward the mirror and back from it would be expected to be $[c^2 - v^2]^{1/2}$, as would be the case for a boat that crosses the stream toward the other side.

It then follows that the times taken for a point on a light wave (say one of its maxima) to return to the starting point would be different when moving in the directions parallel to the Earth's surface and perpendicular to it. Thus, if two waves are in phase when they leave a common source, one propagating parallel to the Earth's surface and the other propagating perpendicular to it, they would be out of phase when they return, even though they went along equal pathlengths—because the speeds of the light in the two directions are presumably different. The measured phase difference should then turn out to be dependent on the speed of the Earth relative to the ether.

The result of the Michelson-Morley experiment was negative. That is, no interference fringes were observed. It was as though there was no ether medium there, just as the absence of the stream would imply that the boat should move at the same speed in all directions.

To explain the negative result of the Michelson-Morley experiment it may be surmised that when the Earth moves through the ether it 'drags' a layer of ether with it such that there would be no relative velocity between the Earth and the ether. In this case, the speed of light, c, would remain c in all directions, irrespective of the fact that the source of the light is in motion. But even with this idea, it must still be admitted that the ether is at rest at some elevation above the Earth's surface. That is, the ether of the entire universe is not being 'dragged' by the Earth's rotation. Every other explanation proposed to interpret the null result of the Michelson-Morley experiment in the years that followed, was either inconclusive (at best) or logically and scientifically inconsistent (at worst).

One of the most interesting of these attempts was that of Hendrik Antoon Lorentz (1853–1928). He suggested that the null result was due to a physical interaction between the ether and the measuring instrument—an interaction that must necessarily be anisotropic (different in different directions). It was Lorentz's speculation that this newly found interaction caused the lengths of all mesuring instruments to shorten by a factor $[1 - (v/c)^2]^{1/2}$ in the direction of motion relative to the instrument and the Earth, but to be unaffected in the direction perpendicular to the direction of relative motion.

With this idea, if a meter stick should move in the direction of its length at 1.8×10^8 m/s (i.e. at $0.6c$), then its physical length would contract to:

$$[1 - (0.6)^2]^{1/2} \text{ (100 cm)} = 80 \text{ cm}$$

while the width and thickness of the meter stick would remain unchanged. Or, a space-traveler who travels close to the speed of light, say at $v = 0.9999999c$, in the direction of the length of his body, would shorten so much that he would have essentially a two-dimensional body, flattened into a wafer along his height, but with the same dimensions perpendicular to his height. On the other hand, if he should position himself perpendicular to his direction of motion, he would become so thin to an outside observer, he would have to turn sideways to be seen. A further conclusion is that as soon as he returns to Earth, in the reference frame of his observer, he must suddenly regain his original proportions.

The explanation of the null result of Michelson and Morley, by Lorentz's conjecture regarding an anisotropic interaction between the ether and the material of an object that moves through it, was too *ad hoc* for many scientists to accept. Additionally, no explanation was provided for either the physical interaction between the ether medium and ordinary matter as constitutes the measuring instruments, or why it is that this interaction should be anisotropic, other than fitting the data of the interferometer experiment.

As it turned out, in the early years of the twentieth century there was a different interpretation provided for the Lorentz contraction factor, $[1 - (v/c)^2]^{1/2}$, as having to do with a *scale change,* when *expressing* the laws of nature in different reference frames of relative motion with the use of spatial measures. This was in contrast with Lorentz's meaning in terms of a physical interaction that would actually contract the size of a meter stick or a human being who moves relative to some observer. This new interpretation came with one of the two major revolutions in modern physics—Einstein's theory of relativity.

Mercury's Anomalous Orbit

In addition to the apparent anomaly in the behavior of light—propagating throughout space without the need of a medium to conduct it—there was another anomaly discovered pertinent to the development of relativity theory. This was the observation that the smallest planet of our solar system, Mercury, does not return to an equivalent position in its orbit in equal times. It was seen, around the middle of the nineteenth century, that Mercury does not have a precisely periodic orbit. Rather, Urbain Jean Joseph Leverrier (1811–1877) saw that this planet took a small amount of time longer in each of its years to return to the equivalent point on its orbit, relative to the Sun's position. This would be equivalent to the effect of having a rotation of the elliptical axes of the planet's orbit. It is called the 'perihelion precession' of Mercury's orbit because it is the point of closest approach to the Sun's position, occurring at one end of the 'cigar-shaped' orbit) that the astronomers were focusing on. (See figure 8.4.)

At first Mercury's anomalous motion about the sun seemed to defy Newton's theory of gravitation. In accordance with the classical theory, when a 'central force' acts on the planet (as a test body), it must orbit about the source of this force in precisely periodic fashion. However, it was soon realized that the central force of the Sun on Mercury was not the only force acting on it. It was also perturbed by the Newtonian gravitational force of the other sister planets of the solar system that are indeed non-central relative to Mercury's rotation about the Sun's position. Their perturbing effect was calculated toward the end of the nineteenth century, using the knowledge of the positions of all of the other planets of our solar system relative to Mercury; it was found to account for about 90% of the anomalous, non-periodic motion. Yet the astronomical measurements were more accurate than one part in ten. It was then realized that there must be some other source of the remaining 10% of the aperiodic motion.

For a short time it was believed that yet another planet must exist to produce this extra contribution, though located in such a position within our solar system that it would be unobservable most of the time. Some astronomers were so certain about this, because they had so much faith in Newton's law of universal gravitation, that they named the unseen planet, calling it 'Vulcan'. Nevertheless, there was never any corroborating evidence to back up the claim of the existence of 'Vulcan'. It wasn't until the second decade of the twentieth century that the revolution of general relativity theory led to the prediction of this 10% contribution to Mercury's aperiodic motion; only then did this motion became comprehensible.

At this point in the discussion it is important to note that the revolution of relativity theory in the twentieth century replaced the conceptual notion of atomism as a model of matter in physics with the continuous matter field approach, and the positivistic approach to knowledge with that of 'abstract realism'. As we have seen earlier, these two conflicts, one in physics and the other in philosophy, have persisted throughout the history of science. The controversy continued into the twentieth century because of the appearance of two revolutions in physics, *simultaneously*. These were the quantum mechanical approach to knowledge and that of relativity—each taking opposite stands on the questions of atomism versus continuity and on positivism versus realism.

So long as both of these theories would be applied to different domains of phenomena, it did not bother most of the physicists that at the roots of a fusion of the theories there were fundamental dichotomies. But a major problem arose when it was recognized that both of these theories must *necessarily* be fused, for each of them to claim to be logically consistent and complete (chapter ten). With this requirement and in view of the fundamental dichotomies of the two theories, *together,* it is clear that the seeds of a third revolution were sown — eventually to emerge from the present state of modern physics.

In the following chapters we will see that the conflict between the quantum and relativity theories, each claiming to be a fundamental theory of matter, continues into the contemporary period, and then seems to come to a head in our own time, accelerated to a large extent by experimental and theoretical research in high energy physics. In the following chapters we will elaborate further on this conflict, setting the stage for the physics of the future, perhaps not to come to fruition until the twenty-first century.

Three

Early Discoveries About Quantized Matter and Radiation

There were two important experimental investigations in physics that ushered into the twentieth century radically new views of matter and radiation. First there was the study of J.J. Thomson that led to the discovery of the electron, the smallest and lightest of the elementary particles with non-zero mass (a particle whose inertial mass is about 1/2000 of the mass of the lightest atomic element, hydrogen[38]). Second, there was the analysis of 'blackbody radiation' by M. Planck; this is the electromagnetic radiation inside of a metal cavity kept at a constant temperature.[39]

The first of these experimental investigations indicated that not only is matter generally atomistic, down to its smallest unit, but the smallest elements of matter, the electrons, are the carriers of the quantity of electric charge associated with electric current in a metal conductor. The blackbody radiation experiments led to a *partial reversal* of the nineteenth-century discovery that light is strictly a wave phenomenon. For it was discovered that under certain physical conditions, electromagnetic radiation behaved as though it were a collection of discrete particles, while under other sorts of experimental conditions it behaved as though it were indeed a collection of waves. The idea of 'wave-particle dualism' was then proposed by Einstein; who asserted that light is not really a pure wave or a pure particle, rather, it is a wave only if studied under those sorts of experimental circumstances where it would appear as a wave, and it is a particle only if studied under those circumstances where it would appear as particles (as in the 'photoelectric effect' experiment, whereby 'discrete particles' of light, called 'photons,' collide with and release discrete particles of matter, electrons, in creating an electrical current). The concept of 'wave-particle dualism', initiated by Einstein,[40] was the most important precursor for the revolution of quantum mechanics, as we will discuss in detail in succeeding chapters.

These discoveries in the early part of the twentieth century, along with the discovery by Rutherford about the planetary structure of the atom and Bohr's theoretical representation of the discrete atom, set the stage for an entirely new conception of atomic matter, quantum mechanics. This contradicted all the earlier atomistic views of matter as well as general views in the philosophy of physics.

The new 'quantum mechanics', underlying the basic description of atomic matter, also contradicted the ideas of the second revolution in twentieth-century physics—the theory of relativity—a new view of matter also instigated by Einstein. In this chapter we will discuss the early ideas and experiments that led to the conflict between the atomism of quantum mechanics and that of classical physics.

The Discovery of the Electron

Joseph John Thomson (1856–1940) discovered the electron, utilizing an instrument similar to the picture tube of a television set. This is the 'cathode ray tube'. At one end of this tube there is an electron emitter (cathode *E*). This negatively charged matter is then drawn to the positively charged plate (the anode *A*), which has a hole in its center. The electron beam that then emerges from the hole in the anode has a speed that is controlled by the voltage V_o imposed between the cathode and the anode. Assuming that the electron beam, so produced, is indeed made up of a large number of individual 'electrons', each of them would be deflected upward by a perpendicularly oriented electric field between the deflector plates *D*. The precise amount of deflection should then depend on the magnitude of the electron charge, *e,* and its inertial mass, *m.* (It turns out to depend on the ratio e/m). Thomson's experiment then led to a measure of the ratio e/m.

Later on in the twentieth century, the electron charge *e* was measured by Robert Millikan (1868–1953), in his 'oil drop' experiment. This was a matter of ionizing an oil drop by one electronic charge, and exerting an electric field on it that would oppose the gravitational force on it. Thus, with the measure of *e*, according to Millikan's experiment, and the measure of *e/m*, according to Thomson's experiment, the mass *m* of the electron was determined. In the early years of the twentieth century, most of the atomic species of the 'Periodic Chart' had been discovered in chemical research. It was found that the atomic elements are indeed characterized by discrete masses, as Dalton had discovered a century earlier. The lightest of the atoms on the chart was found to be hydrogen. But the mass of the electron was then found to be the order of 2,000 times smaller than that of hydrogen. The other atoms of the Periodic Chart were found to have masses that were approximately integral multiples of the mass of a hydrogen atom; the masses of the atomic species of the Periodic Chart are 'quantized'. This is a fact of nature that, eventually, must be explained from the fundamental laws of physics.

The Discovery of the Atomic Nucleus

Assuming that the size of an atom is proportional to its mass, physicists and chemists at the turn of the twentieth century believed that the atoms of the Periodic Chart must be positively charged, amorphous substances, many times greater in size than the electron, and larger than the hydrogen atom in integral proportions. The hydrogen atom itself was then assumed to be such a substance, with positive electric charge equal in magnitude to the negative electric charge of the electron, thence described by such a spherically charged, amorphous material with the electron moving along its surface, so as to yield no net electric charge for the neutral atom. The other atoms were similarly described with successively greater sizes, in proportion to their greater masses, and increasing numbers of electrons on their surfaces so as to cancel the positive charge of the atoms to yield them as charge-neutral.

This commonly accepted model of the atom was refuted by

the experimental discoveries of Ernest Rutherford (1871–1937), who found that most of the mass of an atom is concentrated in a very tiny fraction of its volume. The constituent electrons of the atom were then seen to be orbiting very far from the nucleus of the atom (where most of the mass is concentrated), at radii something like 10,000 times greater than the radii of the parent nuclei. Thus, Rutherford found that the volume of a typical atom is the order of $(10^4)^3 = 10^{12}$ times greater than the nucleus of this atom, where its mass is stored. Most of the atom's volume is then empty space, just as most of the volume of our solar system is empty space. An implication is that if the empty space of our bodies could be squeezed out of us, we would be around a million-million times smaller. We would be unobservable under the most powerful microscopes available.

The way that Rutherford discovered this surprising result was to perform experiments in which a beam of alpha particles (nuclei of helium, emitted naturally in radioactive decay) is scattered from a target made out of gold foil. In striking the atoms of gold, the alpha particles then scatter from them in a way that must depend on the sizes and geometrical shapes of the gold atom targets. From the distribution of the scattered alpha particles, when they are absorbed somewhere else, it is possible to deduce the detailed size and shape of the gold atoms that did the scattering.

One might compare this method of analysis with the attempt of a blind man to determine the size and shape of a cow by throwing a basket full of tennis balls at it, then taking note of where the tennis balls land on the ground; assuming that the cow does not move while the tennis balls are hitting her and that there is a very large number of balls, it might then be possible to come to a fair estimate of the size and shape of the cow. It was this type of analysis that led to the sizes and shapes of the atomic constituents of crystalline substances, when scattering X-rays from them. But Rutherford's alpha particle scattering led to a size of the atom that was around 10^{12} times smaller than the atomic sizes revealed by X-ray analysis. What he saw was that the alpha particles either passed through the gold atoms or that they scattered backwards from this tiny nuclear volume in their centers.

After Rutherford's discovery, Niels Bohr (1885–1962) proposed the planetary model for the atomic species, with electrons orbiting about the nucleus, under the mutual electric attractive force, similar to the orbiting planets about the sun, under the influence of the mutual gravitational force of the sun and the planets. It was discovered by Newton, in the seventeenth century, that the $1/r^2$ law for the gravitational attraction predicts explicitly the elliptical shape of the planets' orbits, with the center of mass of the sun at one of the elliptical foci (in agreement with Kepler's observations of Mars' orbit, and his further speculation of the universality of the elliptical path for all of the planets). Similarly, the force that binds the electrons to the nucleus, the 'Coulomb force' of electricity, depends on the inverse square of the mutual separation of the nucleus and electrons, $1/r^2$. Except for one major problem, the electric force that would bind the electrons to the atomic nucleus must then also predict that the atomic electrons have elliptical orbits, with the nucleus at one of the elliptical focii.

The problem with this model *in classical physics* is that according to the predictions of the laws of electromagnetism, any electrically charged matter that would move non-uniformly must emit radiation, thereby decreasing its own energy. Since any curved motion is non-uniform, the orbiting electron moves non-uniformly, and according to the laws of electromagnetism it must emit radiation, continually. In this case, the orbiting electron would continually fall in toward the nucleus of the atom. In principle, then, Bohr's theory of atomic structure would not work, since it would predict that no atom, with electrons orbiting about a nucleus, could be stable unless some new principle could be evoked that would nullify the classical prediction of the Maxwell theory. The new principle that Bohr called on, to be operative in the atomic domain, is 'the principle of quantization'. This principle was already used by Planck, prior to Bohr's study of the atom, when the idea was proposed that the energy of electromagnetic radiation is quantized in discrete energy bundles called 'photons'. Bohr's idea for the description of the atom entailed the quantization of the angular momenta of the constituent electrons of atoms. Before discussing Bohr's model in detail, let us first review Planck's discovery of the quanta of radiation.

Max Planck (1858–1947) and Blackbody Radiation

Around the year 1900, a set of experiments appeared on the physics scene that seemed to refute the classical expectations; these experiments had to do with what is called 'blackbody radiation'. The experimental set-up is as follows: a metal box is heated to some fixed temperature, where it is maintained by keeping it in a constant temperature heat bath. While in this *steady state* of constant temperature, the box is emitting and absorbing radiation of all wavelengths; it is then said to be in a state of 'blackbody radiation'. The box may appear to our visual sense as being 'red hot', 'white hot', and so forth. But if we should use filters that would respond to the full spectrum of wavelengths of radiation contained in the box at this temperature, separately, it would be seen that the intensities of the spectral components of this radiation depend on their wavelengths in a particular way.

Thus, by putting a filter in a small window on the side of the heated box maintained at a constant temperature, it would pass only one of the spectral components of the contained radiation to the outside world with a particular intensity. If one should repeat this measurement with different filters for the entire spectrum of contained radiation in the box, and plot their intensities as a function of their wavelengths, one would have the 'blackbody radiation spectral curve'.

Based on classical considerations, and assuming that the radiation in the box is in thermodynamic equilibrium with the heat bath in contact with the box, as described above, physicists Jeans (1877–1946) and Rayleigh (1842–1919) predicted, in the latter part of the nineteenth century, the blackbody radiation curve. The analysis considered the vibrational modes of the electromagnetic radiation in the box, treated statistically as distinguishable 'things', at a particular temperature, in accordance with the statistical mechanical analysis developed by Boltzmann (and independently by Maxwell), as we discussed in the preceding chapter. They correctly predicted the right-hand side of the curve, showing that the intensity values must increase as wavelength decreases toward zero.

Rayleigh and Jeans then predicted that as the wavelengths of the blackbody spectrum tend toward the ultraviolet end of the spectrum (toward very small wavelengths), the intensity of this radiation must tend toward infinity. This effect has been called the 'ultraviolet catastrophe'. Nevertheless, experimentation revealed that the intensity reaches a maximum value when the wavelength decreases to a particular value, and then it starts to decrease toward zero as the wavelength decreases further toward zero—in contradiction to the expectation of the classical theory.

Max Planck was the first to ask the question: what is it that makes this curve peak and then decrease to zero, as the wavelengths are further decreased, in blackbody radiation? In other words, what was wrong in the assumptions made by Rayleigh and Jeans, that led to the ultraviolet catastrophe? Another question that arose from the blackbody radiation curves was: why is it that these curves are exactly the same for metal boxes made out of entirely different materials (such as iron or tin or copper) as was seen in the experimentation around 1900?

The latter question is based on the earlier thinking about electromagnetic radiation, according to Faraday's original interpretation. If radiation is nothing more than a manifestation of the motion of charged matter in the walls of the heated box, then one should expect that the blackbody radiation curves for boxes made out of different types of metals would exhibit correspondingly different spectral curves. This is because the different metallic species (copper, tin, iron, etc.) have correspondingly different electronic configurations in their outer shells. But the observed facts, seen around 1900, were that for a given temperature, the blackbody radiation curves for the different metal cavities were all identical. The implication was that electromagnetic radiation may indeed be a thing on its own, and not merely a representation of the field of force of charged matter in motion.

What Planck recognized was that to derive the experimental blackbody radiation curves when analyzing the average physical properties of a large ensemble of vibrational modes of the electromagnetic radiation inside of the box, as Rayleigh and Jeans had done, the correct shape follows if the further assumption is made that the energy in each of the vibrational modes is linearly proportional to its frequency. That is to say, if ν_1 is the frequency of a particular mode of vibration in the box, and if ν_2 is the frequency of a different mode of vibration in the box, then

$$E_1/E_2 = \nu_1/\nu_2$$

where E_1 and E_2 are the corresponding values of energies in these modes. An equivalent equation to this one is:

$$E = h\nu$$

where the constant of proportionality, h, between the energy of a mode of electromagnetic radiation, E, and its frequency ν, is a universal constant, called 'Planck's constant'. Its value has been found to be

$$h = 6.62517\ (23) \times 10^{-34} \text{ Joule sec.}$$

Thus, in the problem of the metal box, heated to a fixed temperature, the radiation within its walls is in discrete units of energy—'quanta'—that are proportional to their respective frequencies. This discovery by Planck was the initiation of the 'old quantum theory'. Planck's constant, h, plays a significant role in the generalization to atomic structure by Bohr, as we will discuss in the next section, and in the formulation of measurement in the new 'quantum mechanics'.

If electromagnetic radiation is confined to impenetrable walls of a metal box, then *in classical terms* the possible frequency values must lie in a spectrum that is discrete rather than continuous. This follows from the boundary conditions on the solutions of the radiation field equations, that their amplitudes must vanish at the walls. This is the same reason that the frequencies of resonating sound in an organ pipe occurs in a discrete set of modes: the fundamental frequency, ν_o, the first harmonic, $\nu_1 = 2\nu_o$, the second harmonic, $\nu_2 = 3\nu_o$, and so forth. If this is so, *classically,* then what is so surprising about Planck's discovery? It is that since the frequencies lie in a discrete set of values, Planck's quantization condition, $E = h\nu$, implies that so must the energies of these different modes of electromagnetic radiation lie in a discrete spectrum. Thus, certain values of energy in the radiation field in the box do not exist. That is, if E_m and E_n are neighboring energy values for the radiation, there is no other energy *between them* for this radiation. It is in this sense that the total energy for the radiation of the electromagnetic field inside of the box in the blackbody radiation state is a collection of discrete energy 'quanta' (later to be named 'photons').[41]

Planck made the revolutionary leap by then generalizing this way of representing the state of blackbody radiation in terms of discrete quanta to the case where radiation is not necessarily confined to the inside of a box of impenetrable walls. He asserted that electromagnetic radiation, generally, may be considered to be an ensemble of discrete quanta (of photons). The light that reaches us from a distant star is then a beam of such photons, with different energies corresponding to the component frequencies in the beam of white light. When such a beam is sent through a prism, the different colors then pass through at different speeds (dispersion) so as to reveal their separate frequencies in terms of the spectral components of the white light—but now described in terms of a 'particle-like' model, in terms of photons that propagate. It is not precisely the Newtonian corpuscular model, as we discussed in the preceding chapter, but light is nevertheless described as being composed of things with 'particle-like' behavior, as in Newton's model. A fundamental departure from the classical model, however, is that the 'photons' have no inertial mass, while the Newtonian corpuscles of light do have inertial mass. The reason for this difference will be seen later on in the text.

To see that the classical energy concept generally excludes the model of discrete quanta of energy, we may go back to the classical definition of energy as the capability of work being done by the entity that is supposed to have this quality. 'Work', in turn, is generally expressed in terms of the solution of a 'conservation equation'—an equation that predicts that its solutions are constant in time (thus, the law of conservation of energy). In the early part of the twentieth century, it was discovered (by Noether) that if we impose the requirement on all of the laws of nature, in the form of field equations, that their form must not change if the time parameter is arbitrarily shifted by some continuous amount, $t \rightarrow t + \delta t$, then the law of conservation of energy *necessarily* follows. That is, the invariance of the forms of the laws of nature under such *continuous* shifts of the time parameter is a *necessary and sufficient condition* for the incorporation of the law of conservation of energy. When this idea is carried forward to the requirements of the theory of relativity, the fusion of time with space into a continuous space-time language implies that the laws of conservation are a necessary consequence of relativity theory and that 'energy', *per se,* must be a continuously distributed entity. Still, the premiss of the quantum theory rejected this conclusion. As a consequence, energy is not precisely conserved in the quantum theory, as we will discuss later on in the text. If energy is not precisely conserved in the quantum theory, it would then be permissible within its context to allow a discrete energy spectrum in this theory.[42]

According to the quantum theory, when the quantities of energy are the order of magnitude of $h\nu$, the classical notion of continuously distributed energy values breaks down and energy is then described as 'quantized'. But this implies that *in principle,* even in the macro-domain where quantities of energy are much greater than $h\nu$, energy in still quantized—though it is seen as a continuum. It is like looking at a screen door from a distance (it looks like a continuous smear of gray) while up close one sees the discrete grid of the screen. Thus, even in the macro-domain, according to the quantum theory, the seemingly continuous energy spectrum is in reality a 'quasi-continuum'. That is to say, between any two energy values, there is a genuine gap where there is no energy value available. It is just that the gap is too small to worry about in most cases in the macro-domain where measured energies are large compared with $h\nu$. The gap becomes increasingly large as the energies considered become correspondingly small, approaching the order of $h\nu$. This description of an energy spectrum in the macro-domain is to be compared with the actual continuous spectrum in classical physics. In this case, no matter how close two possible energy values may be, there is always room for other energies between them. Such a spectrum, technically, is called 'dense'. It is, metaphorically, like a bus queue in Italy compared with a bus queue in England. In Italy, no matter how close two people waiting in line may be, there is always room for someone else to squeeze between them! But in England, once the line is established, it is against the rules for anyone to break in anywhere! Thus, the Italian queue is 'dense' whereas the English queue is 'quasi-continuous'.

The quantization of light in terms of discrete 'bundles of energy' implies that when a distribution of these quanta, each component having a slightly different frequency than the next, are combined into a given 'wave packet', the mathematical description of this packet shows that it is spatially localized, thereby bringing us back to the corpuscular model of light. But a single frequency component of the light (a 'monochromatic wave', as a yellow spectral component from excited sodium of a sodium lamp) has a mathematical description that has no spatial localization—it is continuously described throughout all space. Thus, the view of a particle of light, called a 'photon', in correspondence to a fixed frequency component, is a peculiar sort of particle in the context of the ancient atomistic model or the atomic model of Newton—for as a particle one would have to say that the 'photon' is everywhere at once. In this case, how could one answer the question: if the photon is everywhere at one time, then where is it going to? The answer to this question does not come until the entirely new theory appears, called 'quantum mechanics', as we will discuss in chapters five and six.

Coming back to the experimental facts about light, it appears that if we examine optical phenomena in one way, for example by focusing light into an interferometer (say the type used in the Michelson-Morley experiment), it manifests itself as a collection of waves—continuously distributed, oscillating fields in space and time that combine in a way that is incompatible with the way discrete particles combine. Yet, under other sorts of experimental conditions, light seems to manifest itself as a collection of discrete particles, exhibiting particle-like trajectories (which would be seen as 'rays') and imparting momentum and energy to absorbing matter in discrete fashion—as though struck by stone pellets. The latter particle-like behavior was seen in experiments that revealed the photoelectric effect, as mentioned earlier.

It must then be asked: what is light, in fundamental terms? If the wave concept and the particle concept are mutually dichotomous from a logical point of view, it seems that light can only be one of these things, even though it may *seem like* the other under particular sorts of experimental conditions. But if it *really* is particle-like, then the appearance of light as being wave-like must only be an illusion, or vice versa!

Nevertheless, the answer to this dilemma that emerged in the early part of the twentieth century, *primarily from Einstein's speculation,* was that light is indeed *both* a wave and a particle, *depending on the way that it is studied in experimentation.* This idea became known as 'wave-particle dualism'. Later on, when the new 'quantum mechanics' was formulated, it was based philosophically on an extension of Einstein's 'wave-particle dualism' by Niels Bohr, called 'the principle of complementarity'—an idea of pluralism in scientific explanation that was extended to matter as well as radiation. Its acceptability as a fundamental feature of matter and radiation, and the subsequent interpretation of the wave aspect of this dualism in terms of the elementarity of probability (an interpretation that was vigorously opposed by Einstein) then led to the revolution of the quantum theory of measurement, known as 'the Copenhagen school' (chapter six). Before getting into the details of this development in modern physics, it is essential to first return to one of its most important precursors: the Bohr model of the atom and the idea of the 'quantum jump'.

Bohr's Model of the Atom

Recall that to explain Rutherford's experimental results on the scattering of alpha particles from atoms of gold, the results that revealed that most of the mass of the atom resides in an extremely tiny fraction of the atom's volume, Bohr postulated that the atom is a miniature solar system. The central (positively charged) nucleus that possesses most of the mass of the atom, plays the role of the central force of the Sun, while the (negatively charged) electrons orbit about the nucleus at distances that are tens of thousands times greater than the nuclear radius (similar to the planets' orbits about the Sun and the relative size of these orbital radii compared with the radius of the Sun).

The major trouble encountered by Bohr's model, as mentioned earlier, is that (in contrast to the gravitational force that binds the planets to the Sun) the electrical force that binds the electrons to the atomic nucleus is not enough to maintain a stable orbit since, according to the laws of electromagnetism, the orbiting electrons lose energy that would radiate away, by virtue of their curved path motions. To resolve the difficulty, Bohr was faced with two alternatives: either abandon the planetary model of the atom and seek a different model to explain Rutherford's data, or keep the planetary model but abandon some aspect of the classical electromagnetic theory so as to maintain stable orbits for the electrons.

Bohr chose the latter alternative. He then took the hint from Planck's discovery that electromagnetic radiation is quantized, postulating that the angular momenta of the electrons in the atomic orbits are also quantized, in units of $h/2\pi$, where h is Planck's constant, originally discovered in the energy-frequency relation, $E = h\nu$. Perhaps this idea of angular momentum quantization was suggested to Bohr because the unit of h is that of angular momentum.

Bohr's explanation of the observed spectra from the analysis of the atom in terms of its discrete energy levels corresponding to each discrete angular momentum of the electron followed only after he postulated further the idea of the 'quantum jump'. This is the idea that if an atom should be in one of its excited energy levels (an energy state greater than its minimum *ground state* level) then the orbital electron must *jump* to a state of lower energy, discretely, continuing this process until it eventually reaches the ground state level. That is, it would generally take several discrete jumps to arrive at minimum energy, but the electron could also jump directly to the minimum energy level.

In order to conserve energy, Bohr saw that when the

quantum jump occurs, lowering the energy of the atom, there must be created, simultaneously, a quantum of radiation, that is a photon, with the same energy that was lost by the de-exciting atom. According to Planck's quantization rule for the radiation produced and the law of conservation of energy, it then follows that the energy lost when the orbiting electron drops from the *n*th to the *n'*th energy level is that created in the photon when the quantum jump occurs:

$$E_n - E_{n'} = h\nu_{nn'}$$

where $\upsilon_{nn'}$ is the particular radiation frequency corresponding to this quantum jump, between the nth and the *n'*th energy level of the atom.

It follows from Bohr's model that if there are an arbitrary number of energy levels between the *n*th and the *n'*th levels, we may have:

$$E_n - E_{n'} = (E_n - E_{n_1}) + (E_{n_1} - E_{n_2}) + \ldots + (E_{n_f} - E_{n'})$$

Dividing this equation by Planck's constant, the following rule is derived for the combination of spectral frequencies of the radiating atoms:

$$\nu_{nn'} = \nu_{nn_1} + \nu_{n_1n_2} + \ldots + \nu_{n_fn'}$$

For example, for a four-energy level system, we have the following frequency combination rule, applied to the radiation spectra emitted:

$$\nu_{41} = \nu_{43} + \nu_{32} + \nu_{21} = \nu_{42} + \nu_{21} = \nu_{43} + \nu_{31}$$

This combination of the frequencies of the lines in the spectrum of emitting atoms was first noticed, empirically, by Ritz, about 8 years before Bohr's analysis. It is called the *Ritz combination principle,*[43] and it fits a very large number of spectra of radiating atoms, thus giving Bohr's model further credence.

Although there was good agreement with the data on the hydrogen spectrum, there were a few deficiencies in the Bohr model. First, the theory was not quite as successful when it was fully applied to non-hydrogenic atoms (helium, berilyum, etc.) Secondly, Bohr's atomic model did not account for all of the spectral lines from a radiating atom—the details of the 'fine structure' of hydrogen were not predicted by Bohr's model. Thirdly, the theory did not account for the 'anomalous Zeeman effect'. This is the appearance of extra spectral lines when the radiating gas is exposed to an external magnetic field.

Finally, serious objections could be raised on the source of the dynamics of the quantum jump, for none was proposed. Rather, it was stated *axiomatically* that an atom in an excited energy level must decay to lower energy levels, though no causal relation was offered that would be responsible for this event. Compare this 'model' with the classical physical situation: a stone sits on the side of a hill, propped against a bush. Taking the bush aside, the stone starts to roll down to the bottom of the hill. The physicists then ask the question: why did the stone roll to the bottom of the hill? The answer given by Galileo was that the Earth acts on the stone, pulling it toward its center. Newton then provided the explicit cause-effect relation—his law of universal gravitation—that predicted the fact that the stone would roll to the bottom of the hill, expressed in precise mathematical language, verifiable with respect to the details of the actual motion observed.

Newton's theory gave a better account of the stone rolling down the hill than did Galileo's theory, which was based on the simple idea that the acceleration of the stone in the direction of the Earth's center is the constant, $g = 32$ ft/s^2, though in this particular example, there is very little empirical difference between Newton's theory of universal gravitation and Galileo's theory of constant acceleration for all bodies, irrespective of their masses. Then, in the twentieth century, Einstein's theory of general relativity superseded Newton's theory. But in all three cases, the effect of the stone rolling down the hill, is *explained* as a consequence of a cause-effect relation, proposed as a law of nature.

On the other hand, in Bohr's theory of the 'quantum jump', there is no cause-effect relation to explain the discrete jump of the orbital electron from the atomic state of higher energy to one of lower energy. The electrical binding of the nucleus to the electron does not predict this quantum jump (as the gravitational binding predicts the stone rolling down the hill). Nor does the quantum jump follow from any other cause-effect relation in his theory. One merely states that the electron jumps from a state of higher energy to one of lower energy, discretely, with the simultaneous creation of a photon of radiation from the radiationless system. Of course, the creation of the photon does satisfy the physical requirement of conservation of energy as it was meant to do. Yet, Bohr does not *explain how* the quantum jump occurs between the atomic energy levels, nor *how* radiation can be created from a radiationless system of matter alone, nor *how* radiation can be annihilated when it is absorbed by other matter. Instead, it is asserted with this model of the quantum jump that an atom in an excited state is *predisposed* to de-excite, simultaneously emitting a created photon with the appropriate amount of energy to satisfy the law of energy conservation.

Such a predisposition of matter and radiation to behave in one way or another, without providing further explanation for this behavior, is not unlike Aristotle's physics, whereby a stone at the top of a hill is said to roll down to the bottom because the bottom of the hill is the more natural place for it to be. With this approach, there is nothing more to say about it.

When the quantum theory of measurement appeared, not too long after Bohr's theory of the atom, the process of the 'quantum jump' and the process of 'photon creation and annihilation', asserted to occur *in acausal fashion,* were described in terms of a new probability calculus, called 'quantum mechanics', which replaced the fundamental cause-effect relations of the classical view, or the field view of Maxwell and Faraday. The laws that determine probabilities were then taken to replace the laws of trajectories of 'things', as the basic laws of nature. Thus it was asserted with this new approach that the laws of nature are indeed laws of chance. And so the end point of a continual reduction toward the most fundamental representation of laws of nature was seen to be a set of relations between matter, radiation and probability.

This view of a fundamental role of probability in science is certainly in contrast with the earlier approaches in which one only *uses* probabilities as a tool, useful where the human mind does not have the facility to calculate all of the variables of the complex system. Nevertheless, it is assumed with the classical viewpoints that the outcome of a physical interaction is *predetermined,* irrespective of the knowledge of a human inquirer, or his instruments. On the other hand, with the new view of laws of matter that emerged in twentieth-century physics, the probabilities become a part of the *fundamental* description of the system of matter and radiation. In this view, an atom in one of its excited states will decay to its ground state (or to other lower energy states along the way) with various degrees of probability. The probabilities for making transitions between these possible energy states of the atomic system then follow from the formal expression of quantum mechanics: the probability calculus that is asserted to be the fundamental law of nature relating to microscopic matter.

In the next chapter we will explore the logic that led to this new interpretation of 'wave-particle dualism', in terms of a 'probability field' that is to accompany the discrete particle of matter. At this stage, however, it should be noted that the flow of ideas in the history of physics was not abrupt at any point. The tension between continuous and atomistic aspects of matter as fundamental has persisted since ancient times. The dispute reached a climax in the nineteenth century, when Faraday's field concept seemed true, *and* the discoveries of the atomistic features of matter by Dalton, Boltzmann and others of the time also seemed to be true. This controversy then seemed to come to a head in the twentieth century, when the experiments of blackbody radiation revealed an apparent need for both the particle and the wave aspect of radiation. Nevertheless, this need had to be further analyzed since these are indeed dichotomous features, when viewed together, under a single explanatory umbrella. The new interpretation of this dichotomy, in terms of 'wave-particle dualism' and thence to the 'principle of complementarity' of Bohr, then led to an entirely new philosophy that carried forward into the twentieth century, to a new underlying theory of matter—an interpretation that remains controversial to this day.

Four

Wave-Particle Dualism and Matter

The de Broglie Hypothesis

The idea of wave-particle dualism, postulated in the first part of the twentieth century by Einstein in order to provide a fundamental explanation for electromagnetic radiation in the micro-domain, was extended by Louis de Broglie (1892–1987) in 1924 to the nature of elementary inertial matter in this domain.[44] In his doctoral thesis de Broglie hypothesized that just as the energy and wavelength of a quantum of radiation are reciprocally related, so the particle and wave aspects of inertial matter, such as electrons, must be reciprocally related. He then postulated that the 'particle' property of an electron, say its momentum p, is related to its wave property, its wavelength λ, as follow:

$$p = h/\lambda$$

where h is Planck's constant—the same universal constant that Planck introduced in the energy-frequency relation for radiation and that Bohr introduced for the angular momentum quantization condition for orbital electrons in atoms.

Noting that h has the same units as angular momentum or mechanical action A (energy $\times$ time), de Broglie recognized that the wave features of matter should become significant only when the mechanical action of a material system is decreased to a microscopic magnitude, of the order of h. One should then be in a position to see the manifestations of matter, such as a beam of electrons, as though they were waves, observing for example their constructive and destructive interference, just as in the behavior of superposition of light waves.

If, indeed, material particles such as electrons have this wave nature, then it follows that if one should focus a beam of electrons with a fixed momentum, and therefore a fixed wavelength, onto a screen S_1 with a small hole in it, one should then observe a diffuse image of the hole on a second screen, S_2, a certain distance away from S_1. This would be in contrast with observing a sharp image, as would be perceived if the electrons were discrete particles. The diffuse image would then make it appear that the electrons are 'bending around the corner' as they propagate past the hole in S_1. This is the well-known 'diffraction effect', discussed in chapter one in regard to the behavior of light waves, an effect that seemed to have ruled out Newton's corpuscular theory of light in favor of the wave theory of Hooke and Huygens back in the seventeeth century. Similarly, the Young double slit experiment, discussed earlier in regard to light, should reveal the same diffraction pattern for an electron beam. (See figure 1.7.)

We understood the meaning of the diffraction maxima and minima in the case of bona fide waves in the classical theory, such as the waves of electromagnetic radiation that follow from the Maxwell equations. But what is the precise *meaning* of these maxima and minima for bona fide particles of matter, such as electrons? That is, how is it that discrete, localized entities can apparently 'cancel each other out' at some places (the diffraction minima) and strengthen each other's numbers at other places (the diffraction maxima)? In particular, what is the *meaning* of the first diffraction maximum, at $\theta = 0°$ in the Young double slit experiment, that reveals a strengthened intensity of the electron beam in the geometrical shadow of the solid portion of the screen S_1, between the two holes?

While these questions were being asked, casting doubt on the de Broglie hypothesis about wave-particle dualism for material particles, experiments were carried out in 1927, revealing, just three years after de Broglie's speculation, that under the proper sort of experimental circumstances a scattered beam of electrons does indeed reveal a diffraction pattern. This very exciting discovery came from the study of C.J. Davisson

(1881–1958) and L.H. Germer (born 1896),[45] and independently by G.P. Thomson (born 1892), on electron scattering from single crystals in which the spacing between the atoms is the order of magnitude of the electron's de Broglie wavelength, $\lambda = h/p$.[46] They found that under these physical circumstances, wherein there is a physical equivalence with the Young double slit experiment, the scattered electrons did fall into a pattern that would be described by diffracting waves. It is interesting that this discovery of the *wave nature* of the electron was made by G.P. Thomson, the son of J.J. Thomson, who thought he had established the *particle* nature of the electron! In this case, the 'generation gap' had to do with the concept of wave-particle dualism.

When the electrons scatter from a crystal lattice, made up of successive layers of atoms separated by a 'lattice constant' of d **cm**, then just as in the case of scattering waves, if the angle of incidence of the electrons is equal to θ, the difference of path lengths of electrons that scatter from one layer of atoms and that from an adjacent layer, d **cm** away, is just equal to $2d \cos\theta$. Now if one observes an interference maximum at a particular angle θ, then the difference of path lengths must be equal to a whole number of wavelengths. (See Fig. 1.6.) The diffraction maxima occur at

$$2d \cos\theta_n = n\lambda \quad (n = 0, 1, 2, \ldots)$$

for the successive scattering angles, $\theta_1, \theta_2, \ldots$

This type of diffraction pattern was already seen in the early part of the twentieth century in observing the scattering of X-rays from the same sort of crystal lattices. But when this pattern is seen in the scattering of supposed particles of matter—electrons—exhibiting that indeed they do have a wave nature, the question we must ask is: precisely what is the *meaning* of the wave nature of the electron? After all, a wave is a non-localized entity (it has a value everywhere in space, in principle) while a particle is supposed to be localized: if it is somewhere at some time, it can be nowhere else then. Similarly, particles add up arithmetically, that is, one electron plus one electron is equal to two electrons and nothing else. But waves add with interference: one wave plus one wave could be a new wave with any amplitude between no waves and two waves. Then what sort of an entity is an electron, if it can be described as a pure wave some of the time and a pure particle at other times and if both descriptions fit the experimental facts?

Schrödinger's Derivation of Wave Mechanics

Erwin Schrödinger (1887–1961) discovered the equation whose solutions were the de Broglie waves. The interpretation of the 'wave function'—the solutions of the 'Schrödinger wave equation'—came later, arousing a great deal of controversy in twentieth-century physics, a debate that is still in progress. But in his original conception of the wave equation, how did Schrödinger himself interpret the solutions of his equation (the de Broglie waves) that had already been empirically verified for material particles, such as electrons?

The answer is that Schrödinger saw his wave equation as a completion of the electromagnetic field equations, providing a more detailed expression for the right hand side that represents the charged matter sources for the electromagnetic field of force. His idea was the following: The charge density function ρ(coulombs/meter^3) is always positive for a positively polarized charge (or always negative for a negatively polarized charge), where 'always' means: at all places and times. But the wave function solution of his equation, the de Broglie wave ψ, is a complex function of the space and time coordinates (rather than a real number function, as the charge density function ρ is), with both positive and negative magnitudes. Indeed, it is necessary for the wave function to be a complex function in order to derive the *interference* effect for the matter of the electron beam. But it is also necessary then to construct the real number function, the charge density ρ, from the complex number wave function. This is done by multiplying the complex function by its complex conjugate. Now the product, $\psi^*\psi$ is always positive, as is the case of the charge density. Thus, Schrödinger said that the generalization of the charge density on the right hand side of Maxwell's equations, is

$$\rho = e\psi^*\psi$$

where e is the charge of the electron, that is, the fundamental electric charge of matter in the sense that all observed charges in nature are integral multiples of e (this is the empirically-confirmed principle of 'charge quantization'). The next step, then, was to find the equation satisfied by the wave function ψ, that was to be interpreted in this way. Schrödinger succeeded in doing this by making an analogy with classical mechanics. Particularly, W.R. Hamilton, a few generations earlier, had shown how to transform the description of optics in terms of rays into the propagation of waves. Starting from this analysis (using the 'Hamilton-Jacobi equation' of classical mechanics) Schrödinger then provided the prescription for determining the de Broglie waves, as the solutions of his wave equation, subject to appropriate boundary conditions.

A Probability Interpretation of the de Broglie Wave

It was the generalization in breaking away from the specific electromagnetic interpretation of the de Broglie wave function that led to a departure from Schrödinger's original meaning. What happened next in the history of wave mechanics was a re-interpretation of this formal mathematical expression to fit the experimental results on the properties of micro-matter in terms of a particular *probability calculus*. The probability interpretation of the wave function became possible as soon as a boundary condition was imposed, asserting that there is no flow of particles described by these waves outside of a particular volume V (a special case could be infinite volume).

By normalizing the wave function, multiplying it by an appropriate number so that the *integral*

$$\int \psi^* \psi \, dV = 1$$

one could then interpret the integrand, $\psi^*\psi$, as the probability per unit volume, that is, a *probability density,* since the unity value for this integral is equivalent to the statement that the total probability (the 'sum' of all possible probabilities) is unity. This is a required property of a probability, in addition to the requirement that each probability can only be a positive number, which is also the property of $\psi^*\psi$.

Thus we may *define* the (always) positive, real number fraction as

$$\psi^*\psi(x,y,z)\ dV = dP(x,y,z)$$

—the probability that a particle would be found at the spatial location (x,y,z), inside of the small (differential) volume element, dV. The integral (sum) of dP over all possible spatial locations would then be equal to one. The procedure then would be to first solve the Schrödinger wave equation for its (complex number) solutions ψ, subject to the boundary condition that ψ automatically vanishes on the boundaries of the volume (supposed to contain the electron). Then substitution into the relation above would give the (differential) probability. The wave function itself is then not a probability; its *modulus*, $\psi^*\psi \equiv |\psi|^2$, is the 'probability density', while the complex function, ψ itself, is called the 'probability amplitude'. The Schrödinger wave equation and the boundary conditions imposed on its solutions then is the form of a particular sort of *probability calculus*. That is to say, the rules of this mathematical formalism correpond to the rules of probabilities of a particular sort. Thus we see how it was easy to interpret Schrödinger's wave mechanics as a probability theory of matter, when correctly describing the wave nature of electrons. Nevertheless, it should be recognized that it is perfectly possible that this mathematical formalism could be some sort of approximation for an entirely different formal expression of the laws of matter in the micro-domain that do not at all obey the rules of a probability calculus. An example of the latter is the theory that Einstein proposed, taking the theory of general relativity to be a fundamental theory of matter. This will be discussed more in chapter ten.[47]

There is a special class of solutions of Schrödinger's wave equation that are called 'stationary states'. These are solutions that may be expressed in a product form, one factor depending only on the spatial coordinates, (x,y,z), and the other factor depending only on the time coordinate, t; that is, the nth solution of this type has the form:

$$\psi_n(x,y,z,t) = u_n(x,y,z)\ e^{-iE_nt/h}$$

The constant E_n appears here as a 'constant of the integration', coming out of the solution of the wave equation. It turns out to be the discrete energy associated with the *n*th stationary state of the system. For example, the orbital states of the atomic problem are such stationary solutions that are associated with the energy levels E_n.

The space-dependent factor in the stationary state solution solves the following type of equation:

$$\hat{O}u_n(x,y,z) = o_n u_n(x,y,z) \quad (n = 0, 1, 2, \ldots)$$

This type of equation is called an 'eigenfunction equation'. The set of *real numbers,* o_n, are called the 'eigenvalues', to be associated with the corresponding 'eigenfunctions', u_n, which are the 'discrete' solutions of these equations. We will see, later on, that the eigenvalues, o_n, are to be associated with the measured values of properties of the observed system, and $\hat{O}$ stands for the measurement operator for this particular physical property. It has the form of instructions in the language of calculus (such as: take the second derivative with respect to the spatial coordinates plus: operate with the function of coordinates $U(x,y,z)$ on the eigenfunction u_n, that appears on their right side.) The prescription for forming the operator $\hat{O}$ is spelled out in the Schrödinger form of wave mechanics.

If one should look upon the discrete set of eigenfunctions, $\{\psi_n\}$ as a 'space of functions' (with a generally infinite number of components, if the index n runs over an infinite range of values), then the constraint on these functions:

$$\int \psi_n^* \psi_n dV = 1$$

is analogous to the Pythagorean theorem in a 3-dimensional space of discrete coordinates (rather than continuously distributed co-ordinates):

$$x_1^2 + x_2^2 + x_3^2 = \text{constant}$$

Such a space of continuous functions, subject to this constraint, is called a 'Hilbert space'. It is the sort of function space that forms a unique mathematical structure for quantum mechanics.

Before going into the Hilbert space formalism of quantum mechanics and its interpretation in physics, we will survey some of the earlier interpretations of this theory, starting again with that of Schrödinger himself; thence to a hydrodynamic interpretation of the wave equation by Madelung, Born's probability interpretation, which follows closely from our discussion above; then to de Broglie's own interpretation of the Schrödinger wave function in terms of his 'deterministic law' of a 'double solution' resolution of the problem of quantum mechanics, in his attempt to restore determinism; and finally to Karl Popper's interpretation in terms of 'propensities' (a form of objective probability) to be associated with the particle of matter.

Five

Early Interpretations of Quantum Mechanics

Schrödinger's Interpretation

Erwin Schrödinger believed that the experimental discovery of the wave nature of the electron, which verified de Broglie's wave-particle hypothesis, was in fact related to a more primitive expression for the electromagnetic variable than normally appears in Maxwell's equations for electromagnetism.[48] These are the charge and current density field variables. That is, the wave nature of the matter fields does not appear explicitly in Maxwell's equations for electromagnetism, but it is implicit in the source terms that give rise to the electric and magnetic fields of force that matter exerts on other electrical matter. Such wave nature then does not reveal itself until the Maxwell field theory is adjoined to wave mechanics. The solutions of wave mechanics—the waves that are to be associated with particles that have electric charge, such as electrons—are then combined with each other to form the (non-wavelike) source fields in the Maxwell equations. But the wave nature of the matter field sources themselves does not reveal itself in experiments that directly test the predictions of Maxwell's electromagnetic equations; it appears, rather, in experiments in the domain of atomic physics such as the scattering of electrons from a crystal lattice, so as to reveal a 'diffraction pattern' of scattered particles.

Schrödinger's investigations were influenced by Planck's earlier discovery of the discontinuity of the *energy exchanged* when light is transferred between interacting atoms of matter. But, along with Planck, he was reluctant to carry the idea of *a state of matter,* as represented by the wave solutions of his equation, to the discrete states of the individual atoms themselves, as Bohr had done. In Schrödinger's view, it was rather like having a continuous distribution of energy states of atoms, *collectively,* though placing a filter between the interacting systems of atoms so that only particular frequency modes would pass from the emitting atoms to the absorbing atoms. The latter transfer of energy was then asserted to occur only when a condition for *resonance* was met. In this view, then, it is the energy and frequency *transfer* between atoms that is discrete, and not the energy values themselves.

Max Planck was also reluctant to accept the idea of the discrete (particle-like) quantum of light. He interpreted his results on the blackbody radiation curve in terms of discrete quantities of *transferred* energies and frequencies, carried by the vibrational modes of the electromagnetic radiation field. Nevertheless, Einstein's use of quanta of light ('photons') to explain the photoelectric effect, Bohr's use of the photon concept in his model of the quantum jump when describing interaction between atoms, and some other early twentieth-century experiments, based on explanations in terms of quanta, were all instrumental in convincing most physicists that the 'photon' was a real elementary particle.

Nevertheless, Schrödinger continued to insist that Einstein's and Bohr's analyses of these data in the atomic domain in terms of discrete quanta, *as things by themselves,* was not a unique explanation since it was actually the *frequency differences* that were involved in the comparisons with the observations. Schrödinger's view then went as follows:

If ν_e is a frequency component of the electronic charge distribution at some initial time, and if ν_e' is its frequency component after this charge distribution has transferred this frequency to some other matter, and if the corresponding frequency components of the absorbing matter are ν_a' (after absorption) and ν_a (before absorption), then the rule that relates the frequency transfer between the coupled emitter and absorber resonators, is:

$$\nu_e - \nu_e' = \nu_a' - \nu_a \qquad (1)$$

Note that the frequencies referred to are not associated with any single particle of a system. They are, rather, the normal modes of vibration of a large many-body system, like the vibrational modes of a large array of spheres, each connected to its nearest neighbor spheres with springs, all arranged with their average positions on the vertices of cubic lattices. Such a system would have its characteristic set of normal vibrational modes, for the whole array, but not for the individual spheres of the lattice.

Equation (1) then says, merely, that the loss in frequency in the emitter portion of the coupled material system must be equal to the gain in frequency of the absorber component. Still, Schrödinger pointed out that for those who wish to believe in the idea of 'quantum jumps' between individual atoms of matter, one merely has to multiply both sides of equation (1) by Planck's constant, h, and then use the definition of the quantized energy, $E = h\nu$, to arrive at the relation having to do with 'energy conservation':

$$E_e - E_e' = E_a' - E_a$$

that is, the loss of energy of the 'emitter' is equal to the gain in energy of the 'absorber'. But the point was that one needn't do this in order to get agreement with the actual data—that in fact entails a transfer of a vibration from some distribution of electrical charge, with density ρ_e, to some other charge distribution, ρ_a, for the respective emitter and absorber components of the coupled system of charged matter.

Schrödinger said, next, that the resonance condition for transferred frequencies (1) may include any number of interacting systems, not necessarily only two. Thus the generalized resonance condition is the following:

$$\nu_e - \nu_e' = \nu_{a_1}' - \nu_{a_1} + \nu_{a_2}' - \nu_{a_2} + \ldots$$

This frequency condition for resonance may then be re-written in the form:

$$(2) \qquad \nu_t \equiv \nu_e + \nu_{a_1} + \nu_{a_2} + \ldots = \nu_e' + \nu_{a_1}' + \nu_{a_2}' + \ldots$$

That is to say, there is a sort of 'total frequency', ν_t, that is conserved when a transfer occurs from the normal modes of vibration of one material system (the 'emitter') to the normal modes of vibration of another (the 'absorber')—*in accordance with a resonance condition.*

This is a familiar situation in classical physics, in the description of coupled oscillators. If they are relatively weakly coupled to each other, but each entails a stiff spring, then if they are initially in a state whereby one of the coupled systems is oscillating with maximum amplitude and the other is at rest, their mutual coupling will have the effect of damping the oscillation of the first part while building up the oscillation of the second part of the coupled system, until the latter part builds up to the same amplitude oscillation that the first part had initially, and the first part has come to rest. The process then turns around, gradually returning vibration to the first part of the system, while the second part dampens to zero, and so on cyclically.

Multiplying both sides of equation (2) by Planck's constant, h, then expresses the law of conservation of energy for the complex system. Nevertheless, as before, this equation arises from a resonance condition on the normal frequency modes of the coupled system, rather than from the law of conservation of the sum of energies of the individual atoms of the complex system.

If Schrödinger was indeed correct about his interpretation of the quantum conditions, then why is it that we see in experimentation that the absorbing material responds *immediately* upon reception of the electromagnetic signal from the emitting material? That is, if this process of transfer is really a sort of classical transfer of vibrations between coupled oscillators, why should we not see a relatively slow build-up of the vibration of the absorbing material, as in the example described in the preceding paragraph? For it is the immediate and localized absorption that seems to lead to the particle-like interpretation of the electromagnetic signal in terms of the 'photon'.

In actual fact in the experimentation of atomic physics, one does *not* see an immediate change in the absorber from the state of no vibration to a state of maximum vibration. One does see a finite time of build-up of the vibration which is measured in terms of the 'line-width' of the absorption spectral line. It is impossible to eliminate this line-width, though it may be made as narrow as one pleases by altering external conditions on the absorbing atoms, such as the pressure (as a collective gas), the temperature, and so forth. The discrete line corresponding to zero line width is an *ideal limit* that is in accordance with a particular model (such as the atomic model in quantum mechanics), but it is not a limit in the model proposed by Schrödinger. And even in the quantum mechanical model, as we will see in the next chapter, the prediction follows that it is impossible to detect a zero line width (immediate response of the absorber) because of an intrinsic feature of the quantum theory called the 'Heisenberg uncertainty principle'.

If Schrödinger's model of an atomic system in terms of *collective* vibrations equally predicts the data on the emission and absorption of electromagnetic radiation, there is no reason at this stage to rule out this model in favor of the quantum mechanical interpretation in a conclusive way. Note once again that the important difference between these models has to do with the atomicity of matter. With the 'photon' model of energy transfer between electrically charged material components of a system, one may imagine that the system is composed of discrete atoms with quantized energy levels, as Bohr originally suggested. But with Schrödinger's model, it is the entire charge distribution of an ensemble that is involved. The latter entails a whole spectrum of frequency components that is available to the incoming radiation. Thus, Schrödinger considered that it would be illusory to think of the data as representing the immediate absorption of 'photons' by the collection of absorbing atoms, or to consider a *single* photon to be absorbed by a *single* atom of matter.

Thus we see that Schrödinger denied that the data of atomic spectroscopy *compels* one to interpret these data in terms of discrete quantum jumps in which the energy of one atom is said to drop suddenly and acausally, thence 'creating' radiation (a photon) to be absorbed later by another atom, which in turn, jumps up to a higher energy level as it 'annihilates' the absorbed photon. In his view, in contrast, it is a pure wave phenomenon that is occurring in the atomic emission-absorption process, having to do with the entire ensemble of material atoms going into a state of resonance, due to the (causal) coupling to a second distribution of electrical charge, distributed among a many-atom system. This was the essence of Schrödinger's interpretation of his 'wave function' solution of his wave equation: that it was a *continuous matter field,* expressing most primitively the feature of matter that it is electrically charged, that it does have *collective wave properties* under the proper experimental circumstances, such as the conditions theorized by de Broglie and then seen experimentally in the electron diffraction experiments of G.P. Thomson and of Davisson and Germer.

Summing up, Schrödinger's wave mechanics, to him, was a necessary addition to the Maxwell field theory of electromagnetism, to complete the representation of the electrical properties of matter. The left-side of Maxwell's field equations are particular combinations of rates of change in space and time of the electrical and magnetic fields of force. The matter field sources of these force fields, that appear on the right-hand sides of Maxwell's equations (the charge and current densities) are 'real number variables' that are factorizable into the more primitive (complex number variable) wave functions, ψ and their conjugates ψ^*, which in turn predict the features of matter in the atomic domain, such as the wave nature observed in the electron diffraction experiments. But it is important to note that in Schrödinger's conception, the wave function ψ for the matter field does *not* relate to a single quantity of micromatter (an electron or an atom, etc). It rather relates to an entire ensemble of matter components. The implication here is that there is no primitive atomistic model of charged matter. Its fundamental description is instead in terms of continuous matter fields.

Madelung's Hydrodynamic Interpretation

An early attempt by Madelung to interpret Schrödinger's wave equation was based on the fact that the 'equation of continuity' in wave mechanics and in the electromagnetic field theory have precisely the same form as the equation of continuity in hydrodynamics.[49] This is the equation that expresses the fact that when there is a fluid flow, then if no new material is created along the way, the quantity of matter that flows must be preserved. That is, the amount of fluid that flows into one end of a cyclindrical pipe is equal to the amount that flows out of the other end if no new fluid is created en route in the pipe.

With the comparison of the dynamical equations of hydrodynamics and the wave mechanics of the quantum theory, Madelung showed that the energy of matter predicted by the Schrödinger wave equation is the sum of a kinetic energy term, a potential energy term *and something else*. The latter was called the 'quantum mechanical potential'. It is the same term that appeared in later extensions of quantum mechanics, called 'hidden variable theories' (chapter seven). To this date, there has been no experimental evidence for this new 'quantum mechanical potential'.

The idea of Madelung's hydrodynamic model is that the mass density of microscopic matter, which determines its hydrodynamic properties, is proportional to the electric charge density of this matter. Note, however, that the preceding results were based on the idea of an ideal fluid—a fluid that has no viscosity, no turbulence and no compressibility. If these could be taken into account in a realistic version of the Madelung model, new results might be in better accord with the experimental facts about micromatter.

Born's Probabilistic Interpretation

Max Born (1882–1970) did not accept Schrödinger's interpretation of matter waves as relating to collective modes of an entire ensemble of atoms. Born was one of the chief proponents of a probability interpretation for the de Broglie waves, relating to single particles or atoms, rather than a collection of particles or atoms.[50] Thus, he interpreted the product of the wave function and its conjugate, which is a positive function for all *r*, multiplied by a differential increment of volume,

$$\psi^*(r)\psi(r)d\,V \equiv |\psi(r)|^2 dV$$

as the probability that a particle, such as an electron, would be in a particular *state,* denoted by the wave function ψ, at the place r and in the volume of space dV. He then interpreted the absolute square of the wave function as a 'weighting function' that follows the electron around, like a dark cloud that continually surrounds it, preventing any exact measurement on any of its physical properties. This was similar to the 'phantom fluid' idea that Einstein attributed to the photons of light. Then, as in Einstein's analogue, when a measurement would be made on some property of the electron, 'the cloud would rain', thereby obscuring any precise determination of that particular property of the electron. It was at this stage of the explanation that the quantum mechanical idea of the 'state' of a system first emerged.

According to Born, then, the absolute square, $|\psi(r)|^2$ is a *probability density* (a probability per volume) that a particle of matter could be found in a measurement to be in a particular state of motion. It follows, next, that a general state of a system of elementary particles, before any measurement is made on the property of one of its constituent particles, is a *sum* of all of its possible states,

$$\psi = a_1\psi_1 + a_2\psi_2 + \ldots\ldots = \Sigma_i a_i \psi_i \tag{3}$$

(This is called the 'principle of linear superposition'). In this summation, the coefficients a_i are generally complex numbers, as are the functions ψ_i. According to Born, the probability density for a particle at the point r is then the absolute square:

$$\psi^*\psi = |a_1|^2|\psi|^2 + |a_2|^2|\psi_2|^2 + \ldots + a_1{}^* a_2 \psi_1{}^* \psi_2 + \ldots \tag{4}$$

This summation then consists of squared terms and cross product terms. The interpretation was that the squared coefficients, $|a_i|^2$, represent the probability that the particle to be measured is in the ith state, of all possible states. For the special case in which an electron, say, would be known to be in the state with $i = n$, it would then follow that $|a_n|^2 = 1$, while all other coefficients with $i \neq n$ would be zero.

With this probability interpretation, what is the meaning of the cross product terms in the summation (4)? In Born's view, the answer is that these are *transition probabilities:* the probability that the electron will jump from one of its states, say the *i*th state, to another of its states, say the *j*th state. This theory of probabilities is then more general than others since it not only entails the probabilities for particular possible states, but it also entails the transition probabilities from each of these states to the others.

There is no dynamical scheme proposed here to explain these 'quantum jumps', just as there was no cause-effect relation underlying the 'quantum jump' in Bohr's original model of the atom. At this stage of quantum mechanics, one merely says that there are quantum jumps that obey a particular calculus of probabilities. The important difference to be noted between Born's and Schrödinger's view of the wave function is that the probabilities defined in the summation (4) are, to Born, in regard to a single microparticle, such as an electron, while to Schrödinger, they refer to an entire ensemble of particles. With Born's interpretation, the fundamental variables of the elements of matter are these probability functions, replacing the velocity and position variables of the Newtonian, classical view.

Born's interpretation fit the experiments dealing with the collisions of particles in predicting the outcomes of *scattering events,* such as

$$A + B \rightarrow A' + B'$$

where (A,B) might be any two particles that scatter from each other, from some initial (unprimed) states of motion to a final (primed) state of motion.

A more general probability interpretation of the state of the system, including transition probabilities, predicts interference effects, thus revealing, for example, the diffraction pattern for electrons scattered from a lattice, or from the two holes of a screen in the Young double-slit experiment. Using the straight probability interpretation, the combined probability of a given electron passing through either of the two holes of the first screen, and hitting the second screen at some location, is the sum of probabilities for going through them:

$$|\psi_1|^2 + |\psi_2|^2 = |\psi_{12}|^2$$

On the other hand, the wave functions for the electron entering hole 1 or hole 2 do not combine in this way, according to the empirical results from electron diffraction. The correct comparison with the data must follow from first adding the wave functions (corresponding to passing through hole 1 or hole 2) and then squaring the sum, rather than first squaring each of the wave functions and then adding them, as above.

With the interpretation of the absolute square of the wave function, $|\psi_{12}(y)|^2$, as the probability for finding the electron at the vertical distance y on the second screen, if either hole 1 *or* hole 2 on the first screen is open *but not both,* we would have:

$$|\psi_{12}(y)|^2 = |\psi_1(y)|^2 \; (or \; |\psi_2(y)|^2)$$

On the other hand, if both holes are open in the first screen, the probability of finding the electron at y on the second screen is:

$$|\psi_{12}|^2 = |\psi_1 + \psi_2|^2 = |\psi_1|^2 + |\psi_2|^2 + \psi_1{}^*\psi_2 + \psi_2{}^*\psi_1$$

(See figure 5.1.)

With this interpretation of the electron wave function, it seemed to the early architects of quantum mechanics, such as Bohr and Heisenberg, that the measuring apparatus must have something to do with the form of the wave function of an elementary particle of matter. In the preceding example, the apparatus is a set of two vertical screens, one of them having two holes that may be opened or closed separately or together. Indeed it was this idea that the measurement had something to do with the nature of the elementary particles of matter that was the truly revolutionary ingredient that was to underlie the quantum theory as a fundamental theory of matter. The fundamental theory of matter then became a particular sort of a theory of measurement: the idea that the variables of matter must entail an observer as well as the observed, though their coupling was to be represented *acausally.* Before coming to the details of this new (nondeterministic) view of elementary matter that came with the new quantum mechanics, we will discuss the early objections to Born's probabilistic interpretation of the elements of matter.

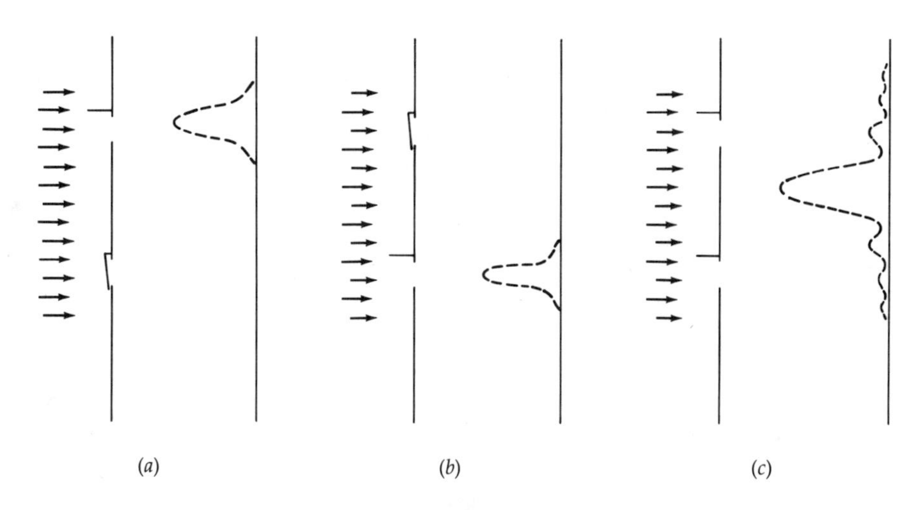

Figure 5.1. Electron Diffraction Studies. *The intensity pattern for electron absorption in each case is: a)* $|\psi_1|^2$, *b)* $|\psi_2|^2$ *and c)* $|\psi_1 + \psi_2|^2 = |\psi_1|^2 + |\psi_2|^2 + \psi_1{}^*\psi_2 + \psi_2{}^*\psi_1$.

Einstein's Objections to Born's Interpretation

Einstein proposed the following *thought experiment* to logically reject Born's interpretation of the linear superposition of wave functions as relating to the superposition of *probability states for a single electron.*

Consider an electron beam hitting a screen with a single hole in it. (See figure 5.2.) Now place a hemispherical film behind the hole, where the hole is at its geometric center. If the hole is small enough, the electron wave would be diffracted rather than moving along a discrete trajectory toward the film. Thus, when the diameter of the hole, *H,* is comparable in magnitude with the electron's de Broglie wavelength, it should behave like a bone fide wave, according to what had been discovered thus far. In this case, in accordance with Huygens' principle (chapter one), the electron, when it arrives at the hole *H,* should manifest itself as a spherical wave, propagating toward the spherical film.

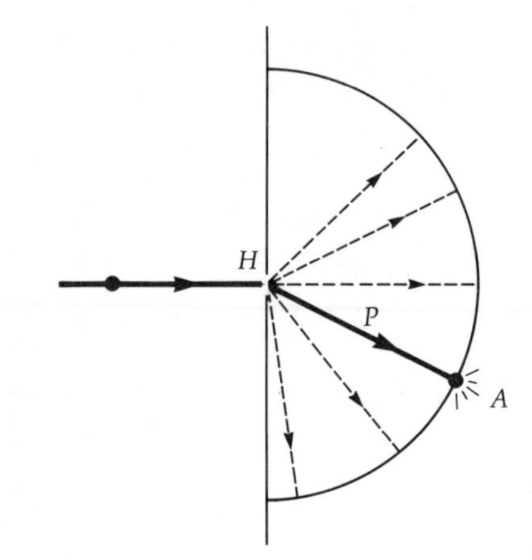

Figure 5.2. Einstein's Thought Experiment—an Objection to Born's Interpretation.

At some later time t after leaving the hole H, a photographic record would be made of the arrival of the electron at some point on the film, say at A. Now according to the probability interpretation, until the electron should be localized at A, *by looking for it there,* it must follow the pattern of the spherical wave function $\psi(r)$. The probability that the electron could be found *anywhere* within the hemisphere, r cm from H, would then be given by the absolute square, $|\psi(r)|^2$. The latter would then be expressible as a linear superposition of all the waves that go to all possible points on the hemispherical film. But as soon as the electron is localized at the particular point on this film, at A, Einstein saw that one must say that the wave function for the electron $\psi(r)$, that is the superposition of all of these waves that go to all of the points of the film (the collection of all such waves forms a 'wave packet') must collapse to the single state function that corresponds to the electron landing at A. This is so because once the electron is seen to be at A, the probability that it is anywhere else on the film is zero. That is, *by looking* for the electron at a particular place on the film, the entire wave packet that describes the electron before it was looked for collapses to the single wave that goes to A.

Einstein felt that this conclusion seemed mysterious. For, by merely looking for the electron at a particular point on the film, one seems to *cause* the electron wave to converge to a point, from all other points on the film. It also implied that as soon as the electron would be seen to arrive at *A*, all other spatial points on the film must respond to this information *spontaneously*. However, according to Einstein's theory of special relativity (to be discussed in detail in chapter eight), there cannot be any spontaneous action-at-a-distance. All signals must propagate at a finite speed, whose maximum value is the speed of light in a vacuum. Thus we see that one of Einstein's main objections to Born's probability hypothesis was that there is no prediction of a causal link (path *P*) that takes the electron from hole *H* to the absorption point on the film, *A*. His further objection was that the Born interpretation implied that there must be spontaneous collapse of a group of waves that are spatially spread out throughout the hemispherical domain in a way that would violate an underlying feature of special relativity theory: the idea that signals may not propagate infinitely fast. That is to say, 'action at a distance' is incompatible with special relativity, but quantum mechanics, according to Born's interpretation, seems to rely implicitly on action at a distance.

Born might have replied that Einstein's claim about a violation of special relativity was based on a faulty interpretation of the wave function, assuming that it is an 'objective' feature of the electron, just as the wave nature of an ocean breaker is an objective feature of the ocean. But Born might have responded that Einstein's objection is removed as soon as one realizes that ψ is really a subjective variable, having to do with an investigator's *knowledge* of the state of the electron. That is to say, before one looks at the film in Einstein's thought experiment, one would say that the electron *could be anywhere* within the hemispherical domain. But after looking at the film and verifying that indeed the electron did land at *A*, the investigator's knowledge about the whereabouts of the electron would become certain. That is, the linear superposition of states of the electron before the observer looks for it, is no more than an expression of the uncertainty in his knowledge regarding where it is. The 'collapse of the wave packet' then only represents the certainty that is gained when the measurement is carried out.

But this response turns Born's original view around, since the electron wave function was originally viewed as an intrinsic probability that is tied to the electron itself (like a gray cloud that continually hovers over it). It was not viewed then as tied to a human inquirer's state of knowledge about the electron. Nevertheless, it was the latter interpretation of the wave function in terms of human knowledge that was to emerge in the meaning attributed to quantum mechanics, as an underlying law of elementary matter. This latter view upholds the positivistic epistemology that was to come into full bloom with the Copenhagen approach to the laws of nature, (as we will discuss in chapter six). Let us now discuss Schrödinger's objection to Born's probability interpretation.

Schrödinger's Objections to Born's Interpretation

Colliding Particles

One of Schrödinger's objections to Born's probability interpretation of the electron wave function was expressed in terms of the following thought experiment: Particle 1 and particle 2 approach each other, where particle 1 is in the state ψ_1 and particle 2 is in the state ψ_2. After an elastic collision they separately go into different states, denoted by ψ_1' and ψ_2', respectively. Now according to the laws of conservation of energy and momentum, and the assumption that there is a single particle meaning attributed to the wave functions, when the scattering of particle 1 occurs, corresponding to the change of wave function, $\psi_1 \rightarrow \psi_1'$, it must follow *unambiguously* that the scattering of the second particle is *predetermined,* and can result in one and only one final state, $\psi_2 \rightarrow \psi_2'$. The precise correlation follows from the conservation of energy and momentum, that the *sum* of the momenta of both particles and the *sum* of the energies of both particles must be the same before and after the collision.

On the other hand, given Born's probability interpretation of the wave function as relating to a single particle of matter, there can be no precise localization. Rather, the particle dynamics must be expressed in terms of wave packets, i.e. a number of possible wave trains, denoted by the possible states of motion for particle 1: ψ_1, ψ_1', ψ_1'', ... Particle 2 may also have an infinite variety of possible states of motion, ψ_2, ψ_2', ψ_2'', ... If observations reveal that particle 1 is definitely scattered to the state ψ_1' (corresponding to particular momentum and energy for that particle) then the laws of momentum and energy conservation *requires* that particle 2 *must* go into state ψ_2' (corresponding to specific momentum and energy for particle 2 that would conform with the conservation laws). This correlation has nothing to do with 'looking at' particle 2, that is, with the act of measurement.

Thus, even if the scattered particle 1 should be located in New York, and particle 2 is located in the vicinity of Tokyo, the observation that reveals that particle 1 goes from the state ψ_1 into the state ψ_1', in the vicinity of New York, tells us that the particle 2 must be in a definite place (Tokyo) with particular values of energy and momentum, without our having to make any measurements on particle 2—so long as it initially scattered from particle 1. Schrödinger then asked this question: what made the collapse happen, whereby particle 2 that was originally in all possible states of motion, suddenly collapsed into one state? Was this caused, Schrödinger asked, by the observer's measurement on the state of motion of particle 1 in New York? That is, did the collapse of the 'wave packet' in Tokyo take place because of an observation that was carried out in New York?

Viewing this possibility as nonsensical, Schrödinger could not accept Born's interpretation of the linear superposition of states as having anything to do with the state of knowledge of a human observer in regard to an observed electron or other sorts of elementary particles or atoms of matter.

The Cat Paradox

Another of Schrödinger's objections was his conclusion that a logical paradox occurs when it is claimed that the concept of 'wave-particle dualism' and the whole formalism of quantum mechanics apply only to the microscopic quantities of matter, while the physical behavior of matter in the macroscopic region must follow from the rules of classical physics. Schrödinger proposed a thought experiment to demonstrate this paradox. It was as follows:

Suppose that a box has a wall in its center, with a hole in it, not much bigger than the de Broglie wavelength of some elementary particle ejected by a radioactive salt. One section of this box contains a radioactive material, such as radium, which is emitting alpha particles at some rate. The other side of the box contains a live cat. On the cat's side of the wall we place a Geiger counter, near the hole in the wall. Connected to this counter is an electrically activated solenoid, connected in turn to a hammer that is aimed at a bottle filled with a poison gas. If the bottle should break, the gas would be emitted and the cat would die. (See Fig. 5.3.)

As soon as the radioactive salt is placed into the compartment next to that of the cat, it emits alpha particles, with associated probabilities. If an alpha particle were emitted with a state of motion that directed it through the hole in the wall, it would activate the instrumentation on the other side of the wall that in turn would kill the cat. If the state of motion of the alpha particle did not direct it through the hole in wall, the cat would remain alive.

Schrödinger then argued that according to Born's view, the

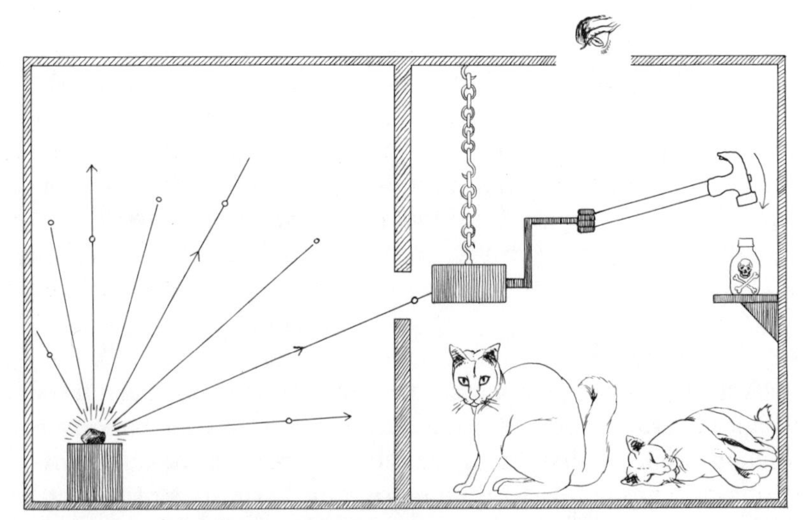

Figure 5.3. Schrödinger's Cat Paradox

alpha particle is in a linear superposition of all possible states of motion, before the measurement is made, that is, before one looks in to see the state of health of the cat. But this must then correlate with a linear superposition of the states of life and death of the cat. Thus, the *macroscopic cat* would, in this case, obey the rules of the quantum theory, in contradiction with the claim of quantum mechanics that only microscopic matter may obey these rules. Then, as soon as one would look into the cat's side of the box, the cat would spontaneously project into the state of life or the state of death. Schrödinger saw this as a paradoxical state of affairs and concluded that the probability interpretation of his wave equation is logically inconsistent, *when applied to the case of a single particle.* Thus he believed that Born's interpretation of quantum mechanics is philosophically and physically indefensible.

De Broglie's Interpretation

It is significant that some of the chief architects of the early form of quantum mechanics and the idea of wave-particle dualism, did not agree with the intrinsic probabilistic interpretation of the matter wave that emerged with the Copenhagen view of quantum mechanics, coming largely from Born's view and thence to the ideas of Bohr and Heisenberg (as we will discuss in the next chapter). One of these architects, de Broglie, postulated (with Einstein, another of the original architects) that indeed there must exist an 'objective' electron, and that it must have a complete description in nature. Thus he noted that the probability calculus that comes with Schrödinger's wave mechanics likely relates only to the 'subjective' aspects of a measurement of the electron's properties—that is, to the chance that an observer would 'see' the particle of matter in one state of motion or another.

To complete the description of the particle, de Broglie then speculated that in addition to the wave function ψ that solves the equations of wave mechanics, there must be another mathematical function that depends on the space and time variables, and entails the dynamics of the *actual trajectory* of the particle of matter, for example, the path in Einstein's thought experiment (figure 5.2) that leads from the hole in the screen *H* to the place on the hemispherical film, *A*, where the electron actually lands. This second mathematical function, in de Broglie's theory, must then *complete* the description of the electron, solving a second equation that complements the equation of wave mechanics. This is called his 'double solution' theory.[51]

Though de Broglie's 'double solution' conjecture is still speculative, he was able to deduce some of the features that this solution must have. First, he saw this solution as describing a point singularity, rather than the continuously distributed feature of the matter wave solution of Schrödinger's equation. He argued that this point singularity must nevertheless be coupled to the Schrödinger wave ψ, which in turn must influence the point singularity function, which we call ξ. Because changes in ξ must influence changes in ψ, and ψ in turn influences ξ, de Broglie concluded that the equation in ξ must be *nonlinear* (as though ξ influences itself!) But he still required that the matter field equation in ψ maintains its linearity, so as to maintain its quantum mechanical form in terms of a probability theory—because this equation is to denote the subjective aspects of the total particle description, that is, the aspects having to do with an observer's *measurements* on the physical properties of the particle of matter. One further reason to maintain this linearity of the matter field equations, de Broglie thought, was Pauli's claim that linearity is required in order to yield the results of the *Pauli exclusion principle.* The latter is at least an empirical fact of nature that is responsible for a great deal of the properties of many-particle systems, such as the ordering of the Periodic Chart among all of the atomic species.

The nonlinear feature of the law that determines de Broglie's second solution ξ was anticipated to lead to a transformation of the discrete spectrum feature of elementary matter that we associate with the wave mechanical solutions (eigenstates) into a *continuous* (though sharply peaked) matter field. It was shown in my research program that general relativity theory, as a fundamental theory of matter (in all domains), also predicts the peaked nature of the properties of matter (i.e. the discrete

Figure 5.4. Schrödinger wonders! Is the cat both dead and alive?

sequence of mass values, charge values of the nuclei, and so on) but only as a linear approximation for a truly nonlinear description of matter.[4] The latter is the prediction of general relativity that *all* of the physical properties of elementary matter have a continuous distribution of values, thereby leading to other physical predictions that would not satisfy de Broglie's model of matter in terms of singular particles. Though Einstein's view of elementary matter, in terms of a continuum matter field, disagrees with de Broglie's view (where matter is made out of discrete bits), they both agreed on the necessity of having a totally objective description of matter at the outset, in order to explain its behavior.

Popper's Propensity Interpretation

The contemporary philosopher, Karl Popper, also does not accept the validity of 'subjectivity' in the probabilistic interpretations of wave mechanics. He addressed the following question: how can one reconcile the *reality* of the particles of matter—as established, for example, in Boltzmann's analysis of gases (according to Popper)—with the probability calculus (quantum mechanics), a formalism that appeared to be necessary in order to explain the data of atomic phenomena?

Popper's answer to this question is to say that the probability aspect of elementary matter is indeed a feature of its objective reality, not merely a subjective aspect of the measurement made by human observers. His view is that each of the particles of matter has an intrinsic 'propensity' to behave in one way or another, depending on environmental conditions (external forces and so forth). When the particular situation arises, the particle then moves along one trajectory or another, with one value of momentum or another, and so on, *depending on its propensity to do so*.[52] The 'propensity' is an Aristotelean potentiality, a 'power' possessed by the bit of matter. The variables that describe the elements of matter in fundamental terms then entail probability functions that are intrinsic features of these elements, replacing the position and velocity variables of matter in the classical view in terms of their discrete trajectories. In Popper's view, the fundamental features of the elements of matter must be objective (i.e. independent of any observer), as it is in Newton's view. He then replaces Newton's objective trajectory variables with his objective propensity variables.

Popper's philosophic position on elementary matter is similar to Boltzmann's.[53] It is, in this author's opinion, the view of 'naive realism'. With this approach, macroscopic matter must be fundamentally atomistic, composed of discrete quantities of units of matter, because matter appears to us as atomistic on the macroscopic level. That is to say, we see individual rocks, trees, planets, galaxies, and so forth. Thus if we extrapolate to the microscopic level, naive realism tells us that the feature of atomicity must be there as well. Additionally, we do not *measure* the properties of matter with infinite precision. There is always finite error involved in any measurement, no matter how small it may be in any particular observation. With naive realism one might then say that in the domain of the elementary particles of matter the use of probabilities in analyzing the errors in our measurements must carry over to the objective features of matter; even when we do not 'look at it', elementary matter must have some 'objective probability' associated with it, which Popper refers to as its 'propensities'.

It should nevertheless be noted that Popper's meaning of 'probability' is not in the same sense as Born's meaning. To Popper, the probability for an event to happen is a measure of the total number of times that the event happens (given identical circumstances each time) out of all possible events. This is called the 'frequency interpretation' of probability. On the other hand, Born's view is that 'probability' is tied to a single event, a single particle of matter. With the frequency interpretation of probability, the chance that a coin will land heads is 1/2 because half of a very large number of flippings of the coin will land heads. But in Born's single event interpretation, the probability of the coin landing heads is 1/2 because of the nature of the single coin flip—that is, the coin has only two sides, and if they are weighted equally each has the same chance for the final state of the coin.

In spite of their differences, both Born and Popper assign a fundamental role to probability in the laws of matter, a view

Figure 5.5. Einstein does not believe that God was playing with dice.

that Einstein rejected when he said that he did not believe that God was playing with dice when He created the universe. Einstein's view of microscopic physics will be discussed further in chapter nine.

Summary

In this chapter, we have discussed several of the interpretations of the de Broglie wave—the 'matter wave'—and the wave mechanical formalism that yields it as a solution to give a rigorous basis for 'wave-particle dualism'.

Schrödinger's own view was that the wave function solution of his wave equation represents a genuine *matter field,* appearing first in physics in the electrically charged matter source terms of Maxwell's field equations for electromagnetism, that is the charge and current density terms that give rise to the fields of electric and magnetic force. He argued that indeed the wave function associated with the charged particle is a factorization of the (real number) charge density into the product of a complex (wave) function and its complex conjugate function. The latter complex functions in turn represent waves, and they appear so in the *wave part* of the wave-particle dualism that characterizes charged particles, such as electrons, when they reveal a typical wavelike behavior, as in the diffraction pattern of electrons scattered from a crystal lattice. It was also Schrödinger's view that this field of matter is characteristic of an entire ensemble of particles of matter rather than a single particle, just as the charge density source of the electromagnetic field characterizes an entire distribution of charged matter rather than a single, discrete charged body.

Madelung's interpretation of Schrödinger's matter wave is in terms of a hydrodynamic field—a field of fluid flow. With the expression of the hydrodynamics of fluid flow applied to the matter fields, and the similar equation of continuity appearing in hydrodynamics as well as in wave mechanics (implying that matter is neither created nor destroyed) Madelung drew a simple analogy which then led him to a new general expression of the matter wave as a complex function whose amplitude and phase both depend on the spatial and temporal coordinates. His equation differs from Schrödinger's in the appearance of a new term that plays the role of a uniquely quantum mechanical type of potential. It is the same sort of term that was discovered many years later by Bohm and his collaborators in their attempt to restore determinism to physics with a 'hidden variable' approach (chapter seven). However, there has never been any experimental confirmation of this innovation of the formal expression of quantum mechanics.

Born's probabilistic interpretation of wave mechanics was the most influential precursor for the *Copenhagen school;* it gave the interpretation that was eventually adopted by the consensus of the physics community. Born assumed that accompanying each elementary particle of matter (as well as the quanta of radiation) there must be a 'cloud of probability'. This entails the chance of measuring particular physical properties of these elements of matter or radiation at any particular time and place. The wave function itself, being a complex variable and also having both positive and negative values of its amplitude, must then be multiplied by its complex conjugate in order to define a bona fide probability, since probabilities must always have positive value (and be less than or equal to one). That is, according to the meaning of 'probability', it would be meaningless to talk about a 'negative probability'. Yet, it was recognized that we needed the complex number 'probability amplitude' in order to explain the observed wave aspects of elementary particles of matter, such as interference phenomena.

De Broglie did not accept the idea that elementary particles of matter are fundamentally continuously distributed fields in themselves, or that they are fully related to probability functions. He conjectured, rather, that there must exist a real, singular particle all the time, and since the probability amplitudes that are the solutions of the wave equation may only provide (at best) an incomplete description of (his assumed) 'predetermined' trajectories of these particles, there must be a second solution to complete the formal theoretical description of the elementary particle that would provide their precise paths in space and time. Such a completion of the description must then yield a deterministic theory for the particles of matter, even though the measurements may not directly relate to the second part of the asserted complete solution.

The idea of de Broglie's 'double solution' resolution of the problem of matter (the problem of restoring determinism to matter physics) has been an important forerunner for the 'hidden variable theories' (chapter seven).

Popper's propensity theory of elementary matter suggests that the probability feature of an elementary particle is one of its 'objective' features. That is to say, the intrinsic propensity of a particle of matter is analogous to a 'power' that is maintained by that matter: it exists whether or not any observer chooses to make a measurement on any of the physical properties of this matter.

Of these interpretations of wave mechanics, only Born's entails the idea of the elementarity of probability in the description of a single particle of matter, and the irreducible connection of this probability with the measurement of a physical property of micromatter made by a macroapparatus. In the following chapter, we will discuss the evolution of this idea to the view of the Copenhagen School, the epistemological view that is the commonly accepted one in physics today. The leaders of this approach are primarily Niels Bohr and Werner Heisenberg, and their epistemologies, though somewhat different, are both based on the approach of logical positivism.

Six

The Copenhagen School

The leading proponents of the Copenhagen School's interpretation of quantum mechanics were Niels Bohr and Werner Heisenberg. As we have discussed in the preceding chapters, Schrödinger's motivation in his formulation of quantum mechanics came from the desire to derive and explain the wave aspects of wave-particle dualism. This was a matter of providing the 'law' whose solutions were the de Broglie waves, agreeing with the experimental verification of the wave aspect of the electron in the works of Davisson and Germer and of G.P. Thomson. One of the stunning successes of Schrödinger's wave mechanics was its prediction of further phenomena in the atomic domain, notably the prediction of the relations between the lines of atomic spectra, data that had been available from the experimental results of spectroscopy for many years. In this regard, Schrödinger's wave mechanics duplicated the success of Bohr's planetary model of the atom, and successfully predicted other features of these spectra that were not in the domain of Bohr's theory.

Werner Heisenberg (1901–1976), unaware of Schrödinger's successes with wave mechanics in the 1920s, discovered (during the same period) a formalism that equally described atomic spectra. It was not at all a continuous wave theory; it was rather a set of rules for combining groups of numbers, called 'matrices', in an algebraic fashion. His theory was called 'matrix mechanics'.

The philosophy that motivated Heisenberg's investigation

was a form of 'positivism', as Heisenberg expressed it in the introduction to his 1925 article that first presented his theory:[54]

> The present paper seeks to establish a basis for theoretical quantum mechanics founded exclusively upon relationships between quantities which in principle are observable.

It was this epistemological view that initiated the approach of the Copenhagen school. It was taken up and elaborated by Bohr, and others who were active in this school, such as Born, Jordan, and Dirac. It is a point of view in the physics of elementary matter that has been maintained in contemporary science until the present time.

Heisenberg's positivistic attitude in his physics reflected Mach's philosophy, as well as that of the *Vienna Circle,* a philosophical group of the early part of the twentieth century that had an important influence in physics. For example, Heisenberg's comments in the introduction to his famous 1925 paper, referred to above, also expressed a difficulty that Heisenberg had with Bohr's planetary model of the atom. For though Bohr's earlier theory was in large part empirically successful, it did entail 'unobservable' features, such as the asserted discrete electron orbits. Thus it was Heisenberg's plan to reformulate a theory of the atom without the need for specifying actually discrete orbits, aside from what can be observed in experimentation.

To reproduce Bohr's successful theoretical results with a theory that only involved 'observable quantities', Heisenberg invented his 'matrix mechanics'. When it was found to be a successful theoretical scheme, it was extended further to all of the other properties of elementary matter. He was overwhelmingly successful in this task, except for the applications where it became necessary to invoke the rules of the theory of special relativity. For, not too long after the initiation of his matrix mechanics, it was recognized (initially by P.A.M. Dirac) that it is *logically necessary* to extend Heisenberg's theory in a way that would satisfy the symmetry requirement of the theory of special relativity (this is the requirement that this law, as any other law in nature, must have the same mathematical form in all possible inertial frames of reference: that is, frames of reference in which the law is expressed, that generally are in constant motion in a straight line, relative to any other inertial frame of reference). This need was also recognized by Schrödinger, when he initially tried to formulate his wave mechanics in a way that would satisfy the rules of special relativity theory.[55] These tasks encountered difficulties that, in fact, have not been overcome to this day. Indeed, the failure, thus far, to fuse the theory of relativity with quantum theory presents us with the major dilemma of twentieth-century physics. These difficulties and possible resolutions will be discussed in more detail in chapter ten.

Not too long after Heisenberg's results were published, Schrödinger showed that his form of wave mechanics and Heisenberg's matrix mechanics were nothing more than two different versions of the same mathematical formalism. That is to say, each of these sets of mathematical rules to represent atomic matter can be transformed into the other without in any way altering the physical conditions assumed.

This equivalence did not make it easy for the believer in either of these theories to accept the uniqueness of the explanation—of wave mechanics, according to Schrödinger, or of matrix mechanics, according to Heisenberg—as based on a continuum wave theory of matter or based on a rule about combining numbers algebraically so as to fit the data about discrete quanta. It is then understandable that Heisenberg was not too pleased to learn that his matrix mechanics was mathematically equivalent to Schrödinger's wave mechanics.

Heisenberg's Matrix Mechanics

Heisenberg originally discovered that certain combinations of discrete numbers, which may be associated with the frequencies of the lines in emission spectra of radiating atoms, can be classified into particular arrangements that obey the rules of a 'matrix algebra'. The discovery came about as follows:

Start with Bohr's relation for the frequency of a photon that is created when an atom de-excites from one of higher energy, E_n, to a lower energy level, E_m, according to:

$$\nu_{nm} = (E_n - E_m)/h \quad \text{Hertz}$$

The wave function for this created photon was then seen to depend on the following 'complex number' amplitude:

$$A_{nm} = exp(2\pi i\nu_{nm}t) \tag{1}$$

where *exp*( ) = *cos*( ) + *i sin*( ) describes the time-dependent part of a propagating wave, that is, the 'photon', in this case, with frequency ν_{nm}.

Heisenberg saw that according to the previously established *Ritz combination principle* in spectroscopy, any particular frequency of a line in a spectrum is the sum of the frequencies of other lines in the same spectrum. For example, one can always find two frequencies in this spectrum that add up to a third:

$$\nu_{nm} = \nu_{nk} + \nu_{km} \tag{2}$$

He then interpreted the photon wave function as relating to the probability of transition between discrete states, so that the 'amplitudes' A_{nm}, B_{nk}, C_{km} are the corresponding probability amplitudes for the creation of the respective photons with frequencies ν_{nm}, ν_{nk} and ν_{km}. Using the rule of multiplication of exponential functions, $exp(a)\ exp(b)\ exp(c) \ldots = exp(a + b + c + \cdots)$, the product rule,

$$exp(2\pi i\nu_{nm}t) = exp(2\pi i\nu_{nk}t)\ exp(2\pi i\nu_{km}t) = exp[2\pi i(\nu_{nk} + \nu_{km})t]$$

implies that the corresponding probability amplitudes obey the rule:

$$A_{nm} = B_{nk}C_{km} \tag{3}$$

Note that the actual probability for the quantum jump to occur between the nth and mth energy level is the absolute square of the probability amplitude, that is: $P_{nm} = |A_{nm}|^2$, which in turn relates to the intensity of the measured line in the spectrum, corresponding to the frequency ν_{nm}.

Heisenberg's next step was to generalize rule (3) by considering an array of all possible transition probability amplitudes, as follows:

$$(4)\quad A_{nm}=B_{n1}\,C_{1m}+B_{n2}C_{2k}+\cdot\cdot=\sum_{\text{all }k}B_{nk}C_{km}=(BC)_{nm}$$

Thus, we may view A as an N-dimensional array of numbers

$$(A)=\begin{pmatrix} A_{11} & A_{12} & \cdots & A_{1N} \\ A_{21} & A_{22} & \cdots & \cdot \\ \cdot & & & \cdot \\ \cdot & & & \cdot \\ A_{N1} & \cdot & \cdots & A_{NN} \end{pmatrix}$$

and A_{nm} denotes the number in the nth column and mth row of the matrix A. According to the rule discovered by Heisenberg, if (B) and (C) are two other matrices of transition probability amplitudes, between the same sets of energy levels, say from 1 to N, then the matrix of A relates to the combination of the matrices of B and of C according to the rule in equation (4), where the symbol Σ stands for a summation over all intermediate states 'k'. Equation (4) may then be expressed: $A = BC$.

After Heisenberg's discovery of this rule of combination of probability amplitudes for the transitions between atomic states, two of his colleagues, Born and Jordan, showed that his rule (4) was simply the rule for combining matrices B and C to give the matrix A, according to the algebra of matrices[56] (discovered several decades earlier by the mathematician, G. Cayley). Thus it appeared to Heisenberg that there was something fundamental in the use of a matrix algebra to express the laws of atomic phenomena.

One of the interesting features about the use of matrix algebra in Heisenberg's discovery about atomic phenomena is the fact that matrices are not commutative under multiplication. That is, if C and B are two elements of a set of matrices that obey the matrix algebra, then their product is another matrix that generally depends on their order of multiplication, i.e. according to the rule given by equation (4), except for special cases,

$$BC \neq CB$$

The physical interpretation of the non-commutativity of

matrices under multiplication was later asserted to relate to the measurement process in atomic phenomena, leading to important consequences, such as the 'Heisenberg uncertainty relations', as we will discuss in more detail later on.

What Heisenberg then showed was that if one of these matrices of numbers, called H, is the measure of the *energy levels* of an atomic system, identified with the diagonal matrix elements, $H_{nn} = E_n$, and if the off-diagonal matrix elements H_{nm} $(n \neq m)$ relate to the transition probabilities between the nth and mth energy levels, then if the matrix of C corresponds to the measured values and transitions between the values of *any other property* of the system (other than energy), it then followed that the difference between HC and CH relates to the *time rate of change* of the matrix elements of C, that is, for the nmth matrix element,

$$(5) \qquad (HC - CH)_{nm} = (ih/2\pi)(\partial C/\partial t)_{nm}$$

where, as usual, h is Planck's constant, and i is the imaginary basis number, $\sqrt{-1}$. The reason for the appearance of h is related to describing phenomena in the quantum domain and the reason for the necessary appearance of the factor i is connected with the requirement of predicting the interference phenomena, which in turn requires the use of 'complex functions' for the state variables of an atomic system.

Equation (5) is called 'Heisenberg's equation of motion'; it forms the basis of his 'matrix mechanics'. His idea in formulating matrix mechanics was then to represent the laws of micromatter as laws that govern discrete chance occurrences. The continuous matter field, conceived by Schrödinger as a basic conceptual feature of matter (its 'wave' aspect) then had no place in Heisenberg's conceptual outlook. Nevertheless, as we have indicated above, in the final analysis the wave formalism of Schrödinger and the matrix formalism of Heisenberg were two physically and mathematically equivalent ways of expressing the same theory, called 'quantum mechanics'. Each made the same successful predictions about atomic matter. At the time when Schrödinger derived this equivalence, there was no way to separate the differing conceptual views of these two scholars as far as the experimental facts were concerned.

It wasn't until a later period in the twentieth century that attempts were made to formulate a mathematically consistent relativistic quantum theory that would be able to fuse quantum mechanics and the theory of relativity. This was done in order to generate a single law of micromatter and radiation. At this time some investigators, such as Dirac, tried to disentangle the Schrödinger and Heisenberg representations of quantum mechanics. But Dirac's study involved the possibility of abandoning the assertion that the state functions are the elements of a Hilbert space, and the subsequent abandonment of the strict probability interpretation.[57] Such attempts to fuse the quantum and relativity theories are still in flux at this writing. They will be discussed in more detail further on in the text. In the next section we will discuss the currently held interpretation, according to the Copenhagen school, for the probability calculus that is called 'quantum mechanics'—that is, the view in terms of a theory of measurement.

The Measurement Interpretation

We have seen in chapter four that the special case of the 'stationary state' for a quantum mechanical system corresponds to the case where the probability amplitudes that are the 'state functions' for the atomic system, factorize into a product of terms, one depending only on the spatial coordinates and the other depending only on the time. The same stationary state occurs in Heisenberg's matrix mechanics as in Schrödinger's wave mechanics. As we saw earlier in Schrödinger's formulation, the stationary state solutions then reduce to the 'eigenfunction' form that solves:

$$\hat{M}\psi_n = m_n\psi_n \tag{6}$$

where ψ_n is the mathematical function that represents the nth discrete state (out of a generally infinite number of discrete states of the system). The 'operator' $\hat{M}$ is made up of a combination of operations that 'act on' the state function ψ_n, mathematically, such as: 'take the first derivative with respect to spatial coordinates, add this to some function of spatial coordinates, and so forth.

In accordance with the probability calculus of quantum mechanics, the full set of state functions $\{\psi_n\}$ are orthonormalized, in the sense that the integral over all space of their products:

$$\int \psi_n^* \psi_m dV = \begin{cases} 1 & \text{if } n = m \\ 0 & \text{if } n \neq m \end{cases}$$

Now, in accord with the mathematical properties of Heisenberg's matrix mechanics, the diagonal elements of the matrices of his formalism correspond to the set of numbers

$$M_{nn} = \int \psi_n^* \hat{M} \psi_n \, dV = m_n$$

and all of the off-diagonal matrix elements $M_{mn} = 0$. These results follow from the preceding two equations. The numbers, m_n, are then the 'eigenvalues' of the operator $\hat{M}$, *by definition*. But it is always possible to use a different set of orthonormal functions, $\{\phi_n\}$, such that (in accordance with the above definitions of the matrix elements, though with respect to this new basis of "eigenfunctions") it follows that $M_{mn} \neq 0$.

It is important in the physical interpretation of quantum mechanics, whether in Schrödinger's wave mechanics or with respect to Heisenberg's matrix mechanics, that the operator $\hat{M}$ must not depend on the function ψ_n itself. The reason for this is the assumption that the micromatter, described by the state function, ψ_n, must not act back on the apparatus (that is, in 'observing' it) in a dynamical way—that is, it is assumed fundamentally that there is no 'recoil' in the observer that, in turn, would alter the state of the observed matter, through the mutual coupling with the observer. This assumption then implies that the basic equations in the wave (state) functions ψ_n are *linear*. That is to say, if $\psi_1, \psi_2, \psi_3, \cdots$ are all solutions of the eigenfunction equation (6), then *any* linear combination of these solutions,

$$\psi = a_1\psi_1 + a_2\psi_2 + a_3\psi_3 + \ldots$$

is also an equally valid solution of the same equation. This imposed restriction in quantum mechanics is called the 'principle of linear superposition'. It is in agreement with the data about the wave nature of micromatter at low energies, such as the low energy scattering and interference of an electron beam that is deflected from a crystal lattice, as in the experimentation of Davisson and Germer, and of Thomson.

It is important to note that the noncommutation of the matrices of Heisenberg's matrix mechanics corresponds (in a one-to-one way) with the noncommutation of the corresponding linear operators:

$$\hat{M} \leftrightarrow M_{mn}$$

This correspondence is indeed the essence of Schrödinger's proof that his wave mechanics, which entails linear operators, corresponds mathematically to Heisenberg's matrix mechanics, both giving a different version of the same formal expression of quantum mechanics. Thus, the question that arises in relation to matrix mechanics is: precisely what is the physical significance of the noncommutation of the operators (and therefore the noncommutation of the corresponding matrices) in quantum mechanics? In answer to this question, Dirac said that the linear operators that appear in quantum mechanics must be interpreted in the following way: the microscopic system is in a state of motion ψ_n, tacitly presuming that there is a set of measurable properties in this particular state, say we call them $(a_n, b_n, \cdots, m_n, \cdots)$. These *simultaneous* properties follow from the similar eigenvalue equations:

$$\hat{A}\psi_n = a_n\psi_n, \; \hat{B}\psi_n = b_n\psi_n, \cdots, \hat{M}\psi_n = m_n\psi_n, \cdots$$

where all of the linear operators above, associated with the same set of state functions ψ_n, are then *commuting*, i.e.

$$(\hat{A},\hat{B}) \equiv \hat{A}\hat{B} - \hat{B}\hat{A} = 0, \; (\hat{B},\hat{C}) \equiv \hat{B}\hat{C} - \hat{C}\hat{B} = 0, \; etc.$$

It was Dirac's view that the set of *commuting operators* $(\hat{A}, \hat{B}, \hat{C}, \cdots, \hat{M}, \cdots)$ represent the 'acts' in which particular macroscopic measuring apparatuses measure the corresponding physical properties of the microscopic system. These are then related only to *simultaneously determinable* properties of the system. The linear operators that represent the other properties of this system that are not simultaneously determinable with those above do not commute with these linear operators. Thus, the matrices corresponding to the latter do not commute with the matrices corresponding with the measurements associated with $(\hat{A}, \hat{B}, \cdots, \hat{M}, \cdots)$.

The measured values of the particular physical properties revealed by the classical apparatus in each of these experiments are then the eigenvalues $(a_n, b_n, \cdots, m_n, \cdots)$, when the system is in its nth state. The linearity of the eigenvalue equations above then reflects the *assumption* that the dynamical state of the measuring apparatus is *not* dependent on the dynamical state of the micromatter that it probes. This is analogous to the assumption that *in principle* there is no recoil in the apparatus due to its interaction with the matter it probes. For if there were such recoil, the state of the micromatter would be altered to a different state (in a nonlinear way).

With this 'measurement' interpretation of the quantum mechanical state functions and the eigenfunction equations that determine them, it follows that if these are fundamental laws of matter, it would not be meaningful to discuss the properties of the micromatter by itself, when it is not being probed by a macroscopic sized measuring apparatus. It also follows, by altering the way in which a particular physical property is to be measured, ever so slightly, say by changing the linear operator for the measurement infinitesimally, that the property would change. Thus, if

$$\hat{A} \rightarrow \hat{A} + \delta\hat{A}$$

where $\delta\hat{A}$ denotes the small change in the set-up of the measuring apparatus, the altered measurement must then yield (sightly) different eigenvalues for this particular property of the micromatter studied, as follows:

$$\hat{A}\psi_n = a_n\psi_n \rightarrow (\hat{A} + \delta\hat{A})\phi_n = (a_n + \delta a_n)\phi_n$$

where δa_n represents the slight alteration of the property a_n, when the system is in the *n*th state, and where the new state functions, 'belonging to' the new (altered) measurement operator, are ϕ_n—these are the new state functions that are to be associated with all of the new measurement operators that *commute* with the new operator $\hat{A} + \delta \hat{A}$, for all of the simultaneously existing physical properties of the system. Note, however, that in accordance with the principle of linear superposition that is imposed by quantum mechanics, each of the eigenfunctions ϕ_n may be expressed as a sum of all of the eigenfunctions ψ_n.

The salient point here is that this slight change in the measurement represents in principle a different sort of observed micromatter, thereby yielding different results about the microsystem of matter. Such an interpretation of the quantum mechanical equations then is a *subjective approach* to the laws of matter, since here the physical properties of matter depend on precisely how they were observed. Such a subjective definition of the fundamental properties of matter was a truly revolutionary proposal in the history of science. In all other theories of matter the act of observing matter to discern its physical properties is not said to influence its intrinsic features. But with the quantum theory, according to the view of the Copenhagen school, the very act of measurement (of a sample of micromatter, determined by a macroapparatus) influences what may be said about that micromatter, *in fundamental terms*. With this approach, there is nothing to say about elementary matter if it does not incorporate the macromeasurement. That is, the idea of an *underlying*, objective reality, independent of an observation, is rejected as a meaningless concept.

The notion of probability enters the theory as follows: if one should not know precisely which of the eigenstates of $\hat{A}$ a microscopic system may be in, before a measurement is carried out, then it must be asserted that the state function of the system is a linear superposition of all possible states:

$$\Psi = \alpha_1\psi_1 + \alpha_2\psi_2 + \cdots = \sum_n \alpha_n\psi_n$$

Thus, in accordance with our earlier discussion, the actual measurement would be described by the linear equation:

$$\hat{A}\Psi = \langle\hat{A}\rangle\Psi$$

where, instead of the discrete eigenvalue a_n of the operator $\hat{A}$, when the system would be in a discrete state ψ_n, we now have the measured value as the 'expectation value' of the measurement, $<\hat{A}>$, which is a *weighted* sum of the eigenvalues a_n, i.e.

$$<\hat{A}> = |\alpha_1|^2 a_1 + |\alpha_2|^2 a_2 + \cdots = \sum_n |\alpha_n|^2 a_n$$

In this summation, the weighting function for the eigenvalue of the *n*th state, $|\alpha_n|^2$, is then interpreted as the probability that the system could be found, in a measurement, in the *n*th state ψ_n. These coefficients indeed satisfy the rule of probabilities, that

$$\sum_n |\alpha_n|^2 = 1$$

Generally, then, the fundamental statements that may be made about elementary matter are confined, according to this theory, to the responses of a macroscopic measuring apparatus—responses that are asserted to be *necessarily* in the form of probability statements. With this view, *the laws of nature are indeed laws of chance.*

We must still answer the question posed earlier: what is the fundamental meaning of the non-commutability of the linear operators in the quantum theory, as a theory of measurement? The answer is implicit in Heisenberg's equation of motion (5). If the linear operator $\hat{C}$ does not commute with the Hamiltonian operator, $\hat{H}$, for the microsystem, then this operator will change *in time* when the property it represents is measured, in such a way that would include the energy measurement at the time of the measurement. What is implied is that, fundamentally, the property corresponding to $\hat{C}$, does not have well defined, *measurable* values, simultaneously with the measured values of its energy. This conclusion follows from the idea that the measurement of $\hat{C}$ *interferes with* other information about the atomic system that can be known accurately, when the measurement of $\hat{C}$ is carried out.

This idea may be described allegorically as follows: suppose that a person's foot is denoted by the state function ψ_f. *Now suppose that the linear operator* $\hat{S}$ denotes the act of putting a sock onto something, and the operator $\hat{B}$ denotes the act of putting a boot onto something. The eigenfunction equation:

$$\hat{S}\psi_f = s\psi_f$$

may then be taken to mean the following: the act of putting a sock onto a foot (the left side of the equation) is equal to a new sort of foot: one with a sock covering it (the right side of the equation). Let us now *further* apply the operator $\hat{B}$ to the preceding equation:

$$\hat{B}(\hat{S}\psi_f) = \hat{B}s\psi_f = s\hat{B}\psi_f = \psi_{f'}$$

The latter equation then represents the act of putting a boot onto a foot that is already covered by a sock, thereby yielding once again a different sort of foot—one that is first covered with a sock (next to the skin), and on top of this, covered with a boot.

Suppose that we now reverse the order of applying these two operations to the foot. We would then obtain the equation, at first,

$$\hat{B}\psi_f = b\psi_f$$

This linear operator equation denotes the formation of a new sort of foot: one that is covered (at the skin) with a boot. Following this with the operation of $\hat{S}$ then gives:

$$\hat{S}(\hat{B}\psi_f) = \hat{S}b\psi_f = b\hat{S}\psi_f = \psi_{f''}$$

The right-hand side of this equation then represents a foot, first covered with a boot (at the skin) and, on top of this, covered with a sock. Clearly, $\psi_{f'} \neq \psi_{f''}$. That is to say, a foot covered by a sock, at the skin, and then covered with a boot (the usual way) is not the same 'state of foot' as a foot first covered with a boot (at the skin) and then with a sock on top of this.

The latter conclusion is symbolized with the linear operator relation in terms of the nonvanishing commutator of the $\hat{S}$ and $\hat{B}$ operators. Symbolyzing the commutator as

$$(\hat{S}\hat{B} - \hat{B}\hat{S}) \equiv (\hat{S},\hat{B})$$

the preceding results are expressed with the linear operator equation:

$$(\hat{S},\hat{B})\psi_f = (\psi_{f'} - \psi_{f''}) \neq 0$$

In quantum mechanics, the nonvanishing commutator is taken to mean that the operators $\hat{S}$ and $\hat{B}$ do not relate to *simultaneously* measurable properties of the micromatter. The reason for this, in the preceding allegory, is that the act of measurement $\hat{S}$ puts the system (the foot) into a state that is different than it was in before the measurement was carried out. The second measurement, $\hat{B}$, puts the system into a still different state. But the latter state of the foot would not follow if the order of 'acts', $\hat{S}$ and $\hat{B}$, was reversed.

The Heisenberg Uncertainty Relations

With the formal structure of quantum mechanics in terms of the 'act of measurement', we must then say that the very fact that the classical apparatus 'observes' the micromatter alters the state of that matter in such a way so as to *interfere* with (other) simultaneous facts about that matter, that one may wish to ascertain. For example, if the state of an electron is ψ, and if the linear operator representing the measurement of the x-component of its momentum is $\hat{p}$ and the operaor representing the act of measuring its position, simultaneously (i.e. at the time when the momentum is the measured value) is $\hat{x}$, then according to the prescription for forming these operators from quantum mechanics, it follows that they do not commute, and that their commutator is proportional to Planck's constant, h, as follows:

$$(\hat{p},\hat{x})\psi \equiv (\hat{p}\hat{x} - \hat{x}\hat{p})\psi = (-ih/2\pi)\psi$$

In accordance with the meaning attributed to the noncommutation of these operators in quantum mechanics, discussed above, the preceding relation implies that the precise momentum (in the x-direction) may not be specified simultaneously with the precise position of the electron along the x-direction. The more accurately that one may carry out a measurement of one of these variables, the less accurately can one know the other because of the *interference* that one of these measurements induces in the other, *according to this theory.*

Compare this approach with the classical Newtonian theory. In the latter view, the momentum and position of the particle of matter (no matter how small) are precisely *predetermined* from the solutions of his equations of motion (given the boundary conditions of initial position and momentum). One does not have to make a measurement in order to determine the precise trajectory. On the other hand, the quantum mechanical trajectory is *not predetermined.* This is because the variables' values depend on the way in which they are measured. In this sense, the quantum mechanical theory of matter is *nondeterministic,* in contrast with the classical deterministic theory of Newton.

With these noncommutation relations between the momentum and position measurement operators in quantum mechanics, Heisenberg then showed that if $\Delta x = [<x^2> - <x>^2]^{1/2}$ is the root-mean-square deviation in the measurement of the position of the particle, where $< \;\; >$ denotes the quantum mechanical expectation value, as discussed earlier, and if the root-mean-square deviation of the measured momentum has a similar expression, then with the commutation relations above,

$$\Delta p \, \Delta x \geq h/4\pi$$

This inequality is referred to as a 'Heisenberg uncertainty relation'.[58] It has been given a variety of interpretations since the 1920s, when it was discovered by Heisenberg. But the most radical interpretation has been that the inequalities of this type are an expression of a basic limitation of objective knowledge, in regard to the world of microscopic matter—electrons, protons, atoms, and so forth. For it follows from such inequalities that the more accurately that one may determine the position of an electron, say, the less accurately can one know its momentum, at the same time. That is, if Δx is arbitrarily small, then $\Delta p \sim h/\Delta x$ is correspondingly large for measurements on micromatter.

The uncertainty relations are also supposed to be a limitation on our knowledge of the electron (or any other elementary system that is governed by quantum mechanics). That is to say, this is not merely a limitation that is due to the human observer's inability to build a sufficiently accurate apparatus, with higher resolution than would be needed to overcome the restrictions on measurement imposed by the uncertainty relations. The idea here is that there are uncertainties in the nature of matter that cannot be eliminated, *in principle.* The axiomatization of the Heisenberg uncertainty relations into a principle of nature was then called 'the Heisenberg uncertainty principle'.

A Thought Experiment

Heisenberg demonstrated his idea about the irreducibility of the uncertainty relations by reference to 'thought experiments', just as other physicists have done before him, such as Galileo (discussed in chapter one). The following is typical of such experiments, as it was discussed by Gamow:[59] Imagine that an atomic-sized athlete stands on the edge of a cliff, throwing an electron away from him toward the sea. Classical physics predicts that the electron's trajectory is the superposition of the downward effect of the Earth's gravitational force and the linear horizontal motion that the athlete imparts to the electron, initially. The resultant path then has a 'parabolic' shape.

One may think of this parabolic path of the electron, as it is thrown off of the cliff, as 'predetermined', independent of any observation of it. But when the attempt is made to see the electron in its path, one must shine light on it and then look for the reflection of the scattered light. The weakest type of light beam that could be used is a single photon (according to the quantum theory). This is a particle of light, with some frequency ν, energy $h\nu$ and momentum $h\nu/c$.

If one should now attempt to 'look at' this electron with the single photon source of light, the photon must be sent to collide with the electrically charged electron, undergoing the 'Compton effect' whereby the photon scatters with less energy and the electron changes its trajectory, with increased energy and momentum. Suppose that the newly created photon is scattered towards a boat, where a fisherman is waiting to determine the position and momentum of the electron.

The law of conservation of momentum implies that after the photon collides with the electron, it will decrease its momentum by the amount:

$$\Delta p = h\nu/c - h\nu'/c = h\Delta\nu/c = h/\lambda$$

corresponding to the increase in the electron's momentum, where λ is the de Broglie wavelength of the electron that is associated with the frequency loss of the colliding photon.

Because of the possible spread in momenta for the scattered

electron, it would land in a corresponding spread of locations, Δx, relative to the precise location of the fisherman who is waiting to measure its properties. The smallest scatter in location that one may expect is the extent of a whole de Broglie wavelength, which would be seen as the separation between adjacent diffraction maxima in the fisherman's absorbing screen. Such a separation would denote the interference of two waves that leave spatial locations separated by the uncertainty in the electron's position, Δx. That is to say, in the most highly resolved type of experimental arrangement, $\Delta x \sim \lambda$. Inserting this result into the preceding equation for the uncertainty in the electron's momentum, we have the relation

$$\Delta p \, \Delta x \sim h$$

It was from the logical conclusions of such 'thought experiments' that Heisenberg concluded that it *necessarily* follows that the more accurately one may measure the momentum of the elementary particle, the less accurately can the position of the particle be known, at the time when the momentum was specified, in accordance with the reciprocal relation, $\Delta x \sim h/\Delta p$. He argued that such a basic limitation on knowledge about elementary matter is then numerically tied to Planck's constant, h.

The momentum and position variables are called 'canonically conjugate'. There are other pairs of canonically conjugate variables of the elements of matter, such as its energy and time measures. This may be derived from the relation of position to momentum as follows: If an electron is not subject to any external potential, then in classical particle physics its total energy would be its kinetic energy, $E = p^2/2m$.

Using the rules of calculus, the corresponding spread of measured energy would then be

$$\Delta E = \Delta(p^2)/2m = (\Delta p)p/m = (\Delta p)v$$

Since the electron's speed may be expressed as $v = \Delta x/\Delta t$, insertion into the preceding equation on the uncertainty in energy measure gives:

$$\Delta E \, \Delta t = \Delta p \, \Delta x = h$$

where the preceding relation between momentum and position

uncertainties was used in order to arrive at the reciprocal relation between the uncertainty in the measure of energy and the uncertainty in the measure of time (when the energy was specified).

Although Heisenberg's uncertainty relations and the general formal expression of quantum mechanics along with the rules that relate its solutions to the empirical observations did have spectacular empirical success in the early stages of contemporary physics, there was still a great deal of controversy on the *meaning* of this theory. These debates were between the proponents of the Copenhagen school, led by Bohr and Heisenberg (though these two scholars, between themselves, did not agree on all points) and a relatively small number of opponents of this view. It is interesting that the great majority of the physics community followed (and continues to do so) the ideas of the Copenhagen school, while among the very small number of opponents were scholars who themselves were instrumental in the structuring of quantum mechanics, in its initial stages. These were Planck, Einstein, Schrödinger and de Broglie, to name the most prominent opponents. It is also significant that the logical arguments that ensued with the modern views of quantum mechanics have not yet been settled.

In the next chapter we will discuss some of the important objections and counter-proposals to the Copenhagen view. In chapter ten we will discuss the major difficulty of fusing the quantum theory and the theory of relativity, a generalization that we will see is essential if the quantum theory is to survive as a bona fide fundamental theory of elementary matter.

Figure 6.1. Early pioneering physicists who participated in seminal debates on quantum mechanics. (Caricatures drawn from a photograph taken at the Fifth Solvay Conference, Brussels, 1927)

E. Schrödinger W. Pauli W. Heisenberg

H.A. Kramers P.A.M. Dirac A.H. Compton L. de Broglie M. Born N. Bohr

M. Planck M. Curie H.A. Lorentz A. Einstein

Seven

Objections to Quantum Mechanics and Counter-Proposals

Einstein's Photon Box Thought Experiment

Einstein proposed the following thought experiment to refute the energy-time uncertainty relation, $\Delta E \, \Delta t \sim h$:

A box is maintained at thermodynamic equilibrium with its emitted radiation at a fixed temperature, as in the blackbody radiation experiment we discussed in chapter three. Suppose that this box is hung from the ceiling by a spring scale, so that its weight may be read at any time (see figure 7.1). The walls of this box are then emitting and absorbing radiation at the same rate, corresponding to the emission and absorption of the same numbers of photons per second. Thus the energy that is lost by the walls in any time interval is regained by the walls in the same time interval so that the total energy of the walls would be constant: they would be in 'equilibrium' with the radiation emitted and absorbed.

But suppose now that a shutter is inserted into one of the walls, which can be opened for any arbitrary time, Δt, that could allow the escape of (at least) one photon—never to return to be re-absorbed by the walls. In this case the matter of the walls would lose the part of their internal energy equal to $(\Delta m)c^2$, where Δm is the inertial mass lost by the wall due to the escape of the radiant energy, and c is the speed of light. (The relation between radiant energy and the energy of inertial matter equal to mc^2 is a prediction of special relativity theory; it will be discussed in more detail in chapter eight.) The lost matter energy, $(\Delta m)c^2$, is then converted into the radiant energy of the single photon, $h\nu$, *minimally*.

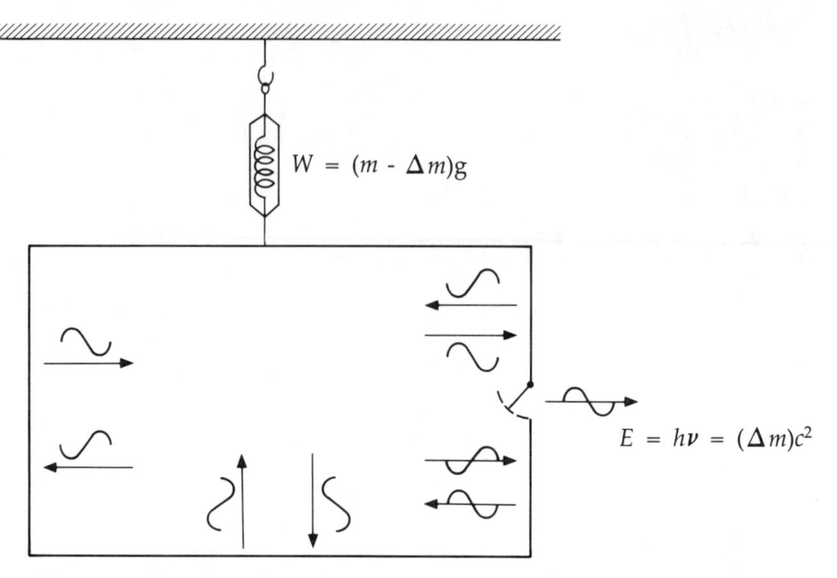

Figure 7.1 Einstein's Photon Box Experiment. *A photon may or may not leave the box during some time interval,* Δt, *when the shutter is open. The weight of the box would then decrease (or not) from mg to* $(m - \Delta m)g$. *This leaves an uncertainty in the internal energy of the box equal to* $\Delta E = (\Delta m)c^2 = h\nu$.

If this photon should not escape during the time Δt when the shutter was held open, the box with total inertial mass m would have maintained its weight, mg Newtons. But if a single photon would have escaped during this time, the weight of the box would have diminished to the amount, $(m - \Delta m)g$ Newtons, which could then be read on the spring scale. The uncertainty in the weight of the box must then correspond to the uncertainty in the internal energy of the box's matter content, corresponding in turn to the energy carried away (or not) by the photon that did or did not leave through the shutter during the time it was open. That is,

$$\Delta E = (\Delta m)c^2 = h\nu \qquad (1)$$

With this thought experiment it is clear that the shorter the duration, Δt, that the shutter is open, the less chance there would be for the photon to escape, thus the more certain one could be about the weight of the box as read on the spring scale.

Einstein then pointed out that the uncertainty in the knowledge of the weight of the box, as determined from a measurement around the time that the photon might leave the box, is not reciprocally related to the uncertainty in the time of measurement; it is rather directly related to the time of measurement. That is, in this example, $\Delta E \propto \Delta t$ rather than $\Delta E \propto 1/\Delta t$. Furthermore, in Einstein's thought experiment of the photon box, the uncertainty in energy and time measurements seem to be totally unrelated to Planck's constant, h. Based on these arguments, Einstein believed that he had refuted Heisenberg's claim that the uncertainty relations are *necessary* to the description of the microdomain of matter.

Bohr's Reply to the Photon Box Experiment

Bohr replied to Einstein's thought experiment by reminding him that he had forgotten to take account of his own discovery of the origin of 'weight' from the theory of general relativity.[60] As we will discuss in more detail in chapter eight, the *principle of equivalence* of general relativity is an asserted equivalence between the description of an accelerated body and a body at rest, though acted upon by an external gravitational field of force. Bohr then argued that in the limit of small time intervals one may replace the accelerated frame of reference (with the acceleration due to gravity, g) with reference frames that move at a constant speed v, thereby allowing the use of the transformations between frames that are derived in special relativity theory (the Lorentz transformations), to be discussed further in chapter eight.

With these assumptions in mind, it turns out that the relative time measures between the different reference frames, that move at a constant speed v relative to each other, is:

$$T' = T[1 - (v/c)^2]^{1/2} \quad (2)$$

where c is the speed of light in a vacuum. If the ratio (v/c) is very small compared with unity, the square root above reduces approximately to $[1 - (v/c)^2/2]$. In this case, the uncertainty of the time measure regarding the different relatively moving reference frames is:

$$(3) \qquad \Delta t = T - T' \simeq T(v/c)^2/2$$

Bohr then continued as follows: according to the principle of equivalence of Einstein's general relativity theory, the energy of a body in motion is *equivalent to* the potential energy that it would have in a gravitational field, when at rest (rather than moving). In the classical approximation, the kinetic energy of a body that moves at a speed v, equal to $mv^2/2$, is then equivalent, according to Einstein's principle of equivalence, to the gravitational potential energy, $m\ \Delta\Phi$. Equating these two terms, one may insert $v^2 = 2\Delta\Phi$ into equation (3) to obtain for the uncertainty in time measure:

$$(4) \qquad \Delta t = (\Delta\Phi/c^2)T$$

Using, now, the position-momentum uncertainty relation that we derived in the preceding chapter, $\Delta p\ \Delta x \sim h$, *that Bohr believed was established earlier, once and for all*, the classical expression of 'force', per se, as a change of momentum with time, gives the following expression for the uncertainty in the measured weight of the box:

$$\Delta W = \Delta p/T = \Delta m\ g$$

It then follows that equation (4) may be expressed as follows:

$$(5) \qquad \Delta t = (\Delta\Phi/c^2)T = (\Delta\Phi/c^2)(\Delta p/\Delta m\ g) = h\Delta\Phi/gc^2\Delta x\Delta m$$

Noting that the weight of the body relates to the gravitational potential as follows: $mg = m(\Delta\Phi/\Delta x)$ and using the energy-mass relation from Einstein's theory of special relativity, $\Delta E = (\Delta m)c^2$, it follows from equation (5) that the Heisenberg uncertainty relation between energy and time,

$$\Delta E\ \Delta t \sim h$$

must hold true, contradicting Einstein's conclusion from his photon box thought experiment.

Based on this rebuttal by Bohr, the physics community believed that Einstein's rejection of the validity of the Heisenberg uncertainty principle had been successfully refuted, thereby establishing, once and for all, the truth of the uncertainty principle.

In answer to this claim, we should note first that the history of science teaches us, most forcefully, that one may *never* claim to have established any truth, *once and for all.* A *scientific theory* is, after all, based on starting axioms which are *always* contingent on the facts of nature and their logical consistency and conformity with other laws of nature that may be brought into their domain. Nevertheless, a certain dogmatic approach did set in, in the 1920s, carrying on during the debates between Bohr and Einstein, into the late 1930s, and to the present time. The idea proposed by many physicists was that quantum mechanics is a mathematical structure that represents the data of the micromatter domain, and because the data cannot be wrong, quantum mechanics cannot be wrong. That is, it has been claimed (by the most orthodox adherents of the Copenhagen school) that quantum mechanics is an irrefutable theory of micromatter.

It is my belief that while the great majority of the physics community has accepted this dogmatic attitude about the meaning and truth value in quantum mechanics, Bohr himself was not that dogmatic about it. One reason for my belief is the fact that Bohr was continually visiting Einstein in Princeton, to discuss their differences, long after the majority of physicists had declared their allegiance to Bohr's view. If he was indeed as confident in his view as were his followers, why did he feel compelled to convince Einstein that he was right in this? Was it, perhaps, that he wasn't completely convinced himself about the Copenhagen approach, and that if he could convince Einstein about it, then he could restore full confidence in it? Of course, this is purely speculative on my part, but, I should like to offer this idea to the reader as an interesting sidelight on this very fascinating story about the evolution of modern physics and its primary players.

A typical response from a contemporary physicist who believes in the Copenhagen view, and has himself contributed greatly to modern physics, is that of R.P. Feynman (1918–1988), who said the following (in his book, *The Feynman Lectures in Physics*):[61]

> The uncertainty principle "protects" quantum mechanics. Heisenberg recognized that if it were possible to measure the momentum and the position simultaneously with a greater accuracy, the quantum mechanics would collapse. So he proposed that it must be impossible. Then people sat down and tried to figure out ways of doing it, and nobody could figure out a way to measure the position of anything—a screen, an electron, a billiard ball, anything—with any greater accuracy. Quantum mechanics maintains its perilous but accurate existence.

I believe that what is fallacious in Feynman's statement is his assumption that experiment actually measures the position or the momentum of an elementary particle (or anything else—a screen, a billiard ball or a planet). What the experimenter does do is to *interpret* a measurement in terms of the position or momentum of a 'thing'. But what he sees is only the reaction of the measuring instrument in terms of a 'signal' that it receives from some other quantity of matter that it explores. It is usually energy and momentum that are transferred between one quantity of matter (the 'measured') and another quantity of matter (the 'measurer'). The interpretation of these responses to signals is then dependent on a *model* that we wish to verify. The model that Feynman tacitly assumes is one of a system of particles of matter. But it is not the only possible model. I will come back to Feynman's position in the next section. First, let us return to Bohr's claim to have refuted Einstein's photon box thought experiment.

In addition to my rejection of the dogmatic attitude of physicists to quantum mechanics (which I do not attribute to Bohr himself), there are other technical points about Bohr's rejoinder that can be argued. The first is in regard to his use of the transformations between the time measures of inertial frames in special relativity. Special relativity applies strictly to the transformations between inertial frames of reference (frames that move at constant velocity in a straight line relative to each other), but Bohr applied these transformations to relatively accelerating frames (i.e., to the *equivalaent of* one frame that falls at the acceleration of $g\ m/s^2$ near the surface of the Earth).

That is, it is illogical in principle to replace the ratio $(v/c)^2$ in the Lorentz transformation formula (2) with the potential of the gravitational field term, $\Delta\Phi/c^2$, as Bohr did to arrive at equation (4), since the potential implies that the equivalent motion would be *nonuniform;* that is, this potential would imply that a body would fall freely at the rate of g m/s^2 *acceleration.*

Practically speaking, the procedure that should be followed in Bohr's analysis, in order to maintain consistency and uniqueness in the predictions, is to *first* derive the comparison of the time measures, T and T' *in general relativity* (i.e. for relatively nonuniformly moving bodies), *then* taking the limit of the transformations as $v(t) \rightarrow$ constant v. But there is no guarantee that this procedure would indeed come close to the formula obtained from assuming at the outset that v is constant (the formula (2)), so long as v remains variable, though coming as close to constancy as one pleases. This is because of the nonlinear feature of the space and time transformations in general relativity theory.

If Bohr's idea about the necessity of incorporating the space-time transformations of relativity theory in order to recover the Heisenberg uncertainty relations is correct, then the problem should be resolvable without the need of any approximations. On the other hand, it is not clear that his analysis would indeed yield the same result if he were to use the general transformations of general relativity without approximation.

Finally, Bohr's use of the reciprocity of uncertainties of position and momentum measures to derive the relation of reciprocity between the energy and time measures seems to me to be circular. For he is putting into his derivation relations that can be expressed in terms of the relations he is trying to prove.

In any case, it is clear that Bohr's alleged refutation of Einstein's photon box thought experiment was not as conclusive as the physics community has claimed it to be.

On Feynman's Defense of the Uncertainty Principle

The salient assumption in Feynman's argument about the fundamental character of the uncertainty principle is that our responses to transferred signals from the 'observed matter' to the 'observing apparatus' is necessarily uniquely in terms of the momenta and positions of particles of matter. That this assumption is false is clear since it presupposes a particular model that is only one out of (perhaps infinitely) many possible models of matter.

One other model that is logically different from the particle model is the one discussed previously in connection with Schrödinger's interpretation of his wave equation. This is a model in which individuation of the elements of matter is *not* one of its basic qualities. Thus, where the transfer of discrete energy quanta between one individial atom and another is the assumed model, it would be out of context with regard to Schrödinger's theory of matter. In his theory, the wave function (the 'de Broglie wave') is supposed to be representative of a continuous *matter wave,* describing a *noncountable* set of elements of matter. This wave description itself is a limit of the reduction toward a most basic representation of matter, in Schrödinger's view. This is a model in which the transfer of quanta is to be understood in terms of a resonance process whereby vibrations are transferred between a part of a large ensemble called 'emitter' and another part of the ensemble called 'absorber'. It is important that, in this view, the transfer of a signal (energy, momentum, and so forth) is not to be interpreted as a single, discrete quantity of energy (a 'photon') that propagates between one single atom and another. As we have discussed in chapter five, Schrödinger claimed that his model could indeed explain quantum phenomena just as well as the single particle model, though without suffering from the logical problems that he saw in the latter (more conventional) view.

Another model in theory that might also explain the data of atomic physics, also not based on the notion of the atomicity of matter, is the conception that emerges from fully exploiting the theory of general relativity, considered as a general theory of matter. For this theory leads to the elementary nature of the continuous field concept in which a matter continuum is a single, inseparable entity. That is, this is an entity that is not a sum of parts, nor is it characterized by a statistical matter field for a non-countable set of entities, as in Schrödinger's view of the 'matter field'. Einstein's view of matter is rather a view in which there are no individual parts in the first place, though it has the asymptotic feature of approaching the *description* of individual parts, in the appropriate (nonrelativistic, linear) limit. It is accurate to use this limit only for computational purposes, when there is sufficiently small energy-momentum transfer between the *modes* of this matter continuum, when they *appear* as though they were indeed separable, individual entities.

The latter picture is analogous to the view of the wholeness of a pond of water, and the asymptotic states of this pond that make it appear to be a sum of parts, such as the individual ripples that glide along its surface. Clearly, while the ripples do appear as separate things at first glance (say, to a grasshopper), they are not really so. That is, the ripples are not separable from the pond as individual 'things', each with a unique weight, size, and so forth. Of course, this is because the ripples of the pond were never 'parts' in the first place. Rather, they were not more than the distinguishable manifestations of the whole pond, and in principle *unlocalized*. One may be able to say where in the space of the surface of the pond a particular ripple peaks, from one time to another, but it is impossible to say where the entire ripple is not, for it is everywhere in the pond, all at once.

The latter is similar to the conception of matter that emerges from the theory of general relativity, as a theory of matter based on the continuous field concept. It is a holistic *philosophy* of matter not unlike the views one reads in various cultural backgrounds, such as the Buddhistic and Taoist cosmologies, the mystical ideas of the Kabbalah, and the arguments of Parmenides, Heraclitus, and Plato.[62] It is a conception that is quite contrary to the atomistic views in Greece, in classical physics or the version according to the Copenhagen school.

For the purposes of this chapter, it is sufficient to say that *both* Schrödinger's view of matter, in terms of continuous matter waves for the representation of an entire ensemble, *and* Einstein's continuous matter theory in general relativity, expressed holistically, logically exclude the atomistic model of Newtonian atomism and the atomism of the Copenhagen school in quantum mechanics. Yet, both Schrödinger's and Einstein's views can explain the Young double slit experiment, without the problems of logical ambiguity that arise in the attempted quantum mechanical interpretation as a manifestation of actual particles giving rise to 'interference phenomena' in the resulting diffraction pattern. There is no such problem in Schrödinger's or Einstein's views because they interpret matter only in terms of fundamental fields, having *only* a wave nature. The modes of behavior of such continuous matter fields then relate to the mutual interactions within this continuum, as a closed system, say between one domain of the interaction that one may wish to associate with 'observer' and another, that one would then associate with 'observed'. Still, there is no actually separable part here. Rather, there are 'modes'—in the same sense that the ripples of a pond or the notes sounded by a violin string are, in reality, the modes of behavior of continuous systems, rather than being 'parts' on their own.

We have emphasized before, however, that it is important that in the limit of sufficiently weak interactions between a 'part' of the system that one may wish to identify with 'observer' and the rest that one may wish to call 'observed' (for convenience in a particular scientific investigation), *it appears as though* there are separate 'parts'. Thus, even in this purely continuum view of matter, there are experimental situations that relate to a description of (almost) particle-like behavior; quanta appear to be discrete manifestations of matter. But this does not entail the concept of *wave-particle dualism,* since *in principle* there is no 'particle' aspect in the first place. There is only 'wave'. Thus, the matter field theories of Schrödinger or Einstein do not entail any dualism in the fundamental explanation of the nature of matter at the elementary level. Instead, the discrete particle aspect of matter is an *appearance* due to physical conditions that prevail in particular experiments where spatial regions of the basic matter continuum are highly concentrated in particular regions, and rarefied in other regions, though still continuous everywhere. This is the model that follows from fully exploiting the theory of general relativity or Schrödinger's matter field theory, even though the limited resolution in the experimentation may not reveal the fullness of the continuous matter field that is in principle there, *underlying* the nature of matter, in any domain.

An example of *localization* of matter is simply the claim to see an individuated chair across the room. Another example is the claimed *localization* of an electron, in an elementary particle collision experiment. Here, the conclusion about its localization may be deduced from the reactions of the instruments, *that are tuned to look for such localization* effects, such as a bubble chamber in a high energy physics experiment. Nevertheless, it is important to recognize that one never actually sees such localization. For example, the 'tracks' that one sees in a bubble chamber are not the particle itself; they are rather the effect of the particle (if it even exists as a discrete entity) on its surroundings. Looking more closely at the matter away from the tracks in the bubble chamber, one sees that these tracks taper off from a central trajectory, but never actually vanish (except at the edge of the instrument) even though they may become extremely weak a small distance from the centrally peaked trajectory. It is analogous to the following situation: suppose that whenever we see the jet trails of high flying aircraft, we never see the aircraft themselves. It might be reasonable then to postulate that there are such individual aircraft that are producing these trails. On the other hand, other models should not be necessarily excluded, such as their production by some strange (to us) sort of interactions within the continuous atmosphere that would give rise to such peaked (though continuous) trails of gas. If the latter is true and not the former, other sorts of experiments would have to be done in order to exclude one of these theories and not the other.

When the distribution of the elementary particles of matter throughout space is truly continuous, though approximated by the description of a truly localized set of 'things', one then arrives at the Heisenberg uncertainty relations in the quantum mechanical description. But within the holistic field theory in relativity, or in Schrödinger's matter field theory, these relations are only an approximation that is mathematically accurate under special conditions; *they are not generally valid.* When we use such an *approximation*, we are ignoring a part of the formal description, for practical purposes of computation. Thus, we only use an incomplete representation of the theory of matter, which we know all the while has a complete representation. This is similar to our use of perturbation methods in celestial mechanics, to estimate the perturbing influence that one planet may have on another. What one does is to first determine the planet's motion due to the Sun alone. Then the added effect on the planet's motion due to the other planets' coupling to it is calculated in terms of a sum of a *convergent series* of terms that depends on the ratio of the masses of the planets to that of the Sun. By cutting off the series somewhere, one obtains an approximation for the motion since the full effect corresponds to the entire series of terms. But since each of the terms in the series is an order of magnitude smaller than the preceding term, it is estimated that this may be an accurate approximation if the series is terminated, say, after the third or fourth term. All the while, however, we know that the series of terms, when it is cut off somewhere, does not give the exact behavior of the planet's motion: it is incomplete. Similarly, from Einstein's field view of matter, the use of quantum mechanics to describe the microdomain is an incomplete approximation, including the use of the Heisenberg uncertainty relations, though this may be *useful* in a particular set of physical circumstances, such as in the representation of low energy transfer phenomena (called 'nonrelativistic physics'). However, from Heisenberg's point of view, calling his uncertainty relations a 'principle', he was claiming that the partial description of micromatter (that Einstein and Schrödinger saw quantum mechanics to be) is all of the theory there can possibly be, in representing this matter.

Heisenberg was led to assert that his uncertainty principle is a basic relation, under all circumstances, because of his wish to maintain the discrete model of matter while at the same time maintaining its (empirically required) wave nature. But in the complete theory that was anticipated by Einstein, one abandons the dualism of wave and particle and replaces this with a single continuum view, implying that one may make use of the uncertainty relations when it is accurate (mathematically) to use a partial description for an actual complete description. For, as we will discuss later on, Einstein's approach yields a logically consistent theory based, once again, on the notion of *predetermination,* though the basic existent in this view of matter is a continuum, rather than a set of individual particles.

The most elementary mathematical variables for the representation of a field theory of matter are continuously variable functions of the space and time coordinates (field variables) rather than the position and momentum variables that describe the trajectories of singular matter particles, as in the Newtonian atomistic view. What is measured in the field view must then be interpreted in terms of the predictions of the field theory, but the fields themselves are not directly related to the observables. That is, according the rules of the field theory, one first solves the field equations for their solutions, then these solutions are used in a prescribed way to calculate particular numbers that are associated with observable effects. It is only in this indirect way that the truth of the field theory is verified or refuted.

The formal expression of the continuum field theory of matter must then asymptotically approach the expression of quantum mechanics, in the low energy (nonrelativistic) approximation. Thus we see that, *in principle,* a continuum field theory would replace the atomistic view of quantum mechanics if the theory of relativity would form a genuine basis for a theory of elementary matter—implying a replacement of the *nondeterministic, positivistic* approach with a *deterministic* approach of a continuum field theory, lodged in the epistemological view of *abstract realism.*

Whether or not the latter view will in the final analysis win out over the quantum mechanical approach, it is clear from its possibility that Heisenberg's interpretation of the data of the atomic domain is *not necessitated,* in expressing the underlying laws of elementary matter. The "perilous but accurate existence of quantum mechanics" that Feynman refers to is then threatened more than he realizes. It is indeed threatened by the possibility of replacement by other theories of matter (that the present day physics community tacitly rejects as impossible), following from an entirely different conceptual standpoint: the stand of the deterministic, continuum theory of matter as is implied, for example, by Einstein's theory of general relativity, as a fundamental theory of matter.

The Role of Variables in Quantum Mechanics

The logical fallacy that I see in the claim of the adherents of the Copenhagen school about the uniqueness of their conclusion concerning a fundamental uncertainty in the description of elementary matter is their tacit assumption that space, time and momentum, associated with the trajectories of elements of matter, are directly perceivable (in the sense of responding directly to their average values) in experimentation. If this were true, there would be some truth in asserting the uniqueness of the use of the atomic model of elementary matter, even though it may not be represented deterministically, as in Newtonian physics. But the fact of the matter is that these are *not* directly perceivable, measured properties of matter. They are, rather, abstractions of a particular model of matter, that are indeed invented for the purpose of providing a useful language to facilitate an expression of a particular underlying theory of matter. It is the theory that is then supposed to provide the *explanation* for the actual data.

As emphasized earlier, the data requires responses within a coupled system of matter, the coupling between the respective domains associated with the 'observer' (the measuring apparatus) and the 'observed' (the micromatter). The space and time parameters are then the 'words' of an invented language that is used for the 'mapping' of the variables that are to explain these data, such as the Schrödinger wave function or the electromagnetic field variables. Technically, the space and time parameters are called the 'independent variables', since they are supplied for the *expression* of the laws of nature, while the field variables that solve the laws of nature, that *depend on* the space and time parameters, are the 'dependent variables': they are the restraints that are in fact dictated by the laws of nature. The 'dependent variables' are imposed by nature; the 'independent variables' are our own invented language, for the sole purpose of *expressing* the laws of nature. The space-time language that we presently use is then perhaps not the most efficient sort of language. Perhaps we will find a better language some day, analogous to Newton's invention of calculus, in order to *express* nonuniform motion.

That is to say, we may be able to exchange the space and time language for a more efficient sort of language, without altering the content of the laws of nature, just as we may transform a sentence from English into French without altering the meaning of that sentence. This, of course, is true so long as the connective relations that are the laws of matter, as expressed with some language system of independent parameters, and their logical relations, are in *one-to-one correspondence* with the connective relations in terms of any other language system. That is to say, the requirement is one of *objectivity,* asserting that any law of nature, if it is indeed a law of nature, must be preserved under transformation to any particular language system. But as far as the language itself is concerned, there are certainly an infinite number of possibilities that could be invented for the purpose of expressing any particular law of nature. As we will discuss further in chapter eight, the objectivity of the law of nature is the fundamental claim of the theory of relativity—it is called 'the principle of relativity'.

An important example of a parameter space other than space-time that may be used to express the law of quantum mechanics in particular, is the energy-momentum parameter language. Because quantum mechanics is a linear theory, the transformation of languages from the ordinary space-time system to the system of energy-momentum parameters is accomplished by taking a 'Fourier transformation' of the equations of this theory, such as the equations of wave mechanics. In this case, the solutions of the transformed equations are *dependent variables,* mapped in a space of energy and momentum variables—the *independent variables.* But the mathematical transformation does not change the physical content of the wave equation in quantum mechanics, nor its full set of physical predictions.

Since we do have at our disposal this possibility of changing the independent variable representation in quantum mechanics, it is possible that difficulties in the mathematical expression of the theory could be overcome by properly generalizing the underlying parameter space—the set of independent variables and their logic. Indeed, this was the idea of the 'hidden variable' theories, where one *adds to* the four-dimensional space-time some additional independent variables, to be used in the mapping of the dependent variables that are the state functions of quantum mechanics. In turn, this is the idea of enriching the set of independent variables while leaving the same degrees of freedom and linear character in the dependent variables—the solutions of the laws of micromatter themselves. The original goal in these studies was to restore determinism to quantum mechanics by *completing* its expression in this way. Before discussing this idea in more detail, however, it should be instructive to go into one more thought experiment, proposed by Einstein and his collaborators, B. Podolsky and N. Rosen. They wrote an article that had a strong influence in the later attempts to construct a hidden variable theory. The latter were attempts to abandon the probabilistic Copenhagen interpretation of quantum mechanics, while still holding onto its formal structure (in terms of the Schrödinger or the Heisenberg representation) though this time trying to *complete* a deterministic description of the elements of matter. These studies grew out of the conclusions of Einstein, Podolsky, and Rosen (referred to hereafter as EPR) that, on the one hand, there must exist a complete description of elementary matter (independent of measurements), but on the other hand, quantum mechanics itself does not provide such a complete description.

The Einstein-Podolsky-Rosen Paradox

While many of the adherents of the Copenhagen school primarily refer to Bohr's refutation of Einstein's *photon box thought experiment* to support their position, the more serious of Einstein's objections, because it came closer to the root of the Copenhagen philosophy, was the so-called EPR paradox. This thought experiment comes closer to the fundamental role of measurement in quantum mechanics, according to the Copenhagen school, and it questions its logical consistency.[63]

The EPR thought experiment was equivalent to the following: consider a two-body system, such as the hydrogen molecule H_2. While each of the hydrogen atoms of this molecule are bound together, its total 'spin state', as measured by its observed angular momentum, is equal to zero. (This is called an '*S*-state' in spectroscopy). Each of the hydrogen atoms has a total angular momentum equal to $h/4\pi$, corresponding to 'spin one-half'). To yield a total spin of zero, the spins of each of the hydrogen atoms in H_2 must then be antiparallel to each other.

Suppose now that some *spin-independent* force is applied to the hydrogen molecule, so as to break it apart, and the individual hydrogen atoms are removed from each other to some arbitrarily large distance. Because the external force that separated them was *spin-independent,* the original spin correlation between these two atoms, that they were antiparallel to each other, must then be maintained. That is, if one of these atoms is found to have a particular spin orientation, say in the 'up' direction (relative to some vertically applied magnetic field), then the orientation of the other atom's spin must necessarily be 'down' relative to it, no matter where it is. For example, if an experiment is performed in New York, measuring the atom's spin to be 'up', then we should be able to specify that, necessarily, the other atom, even if it is in London, should be in the spin 'down' orientation—whether or not we should perform any measurement on the latter—where 'down' means opposite from the 'up' of the first atom.

The claim of the Copenhagen interpretation of quantum mechanics that was challenged by the EPR thought experiment was this: all of the canonical variables of an atom of matter (momentum/position, energy/time, and so forth) are not precisely prescribed, simultaneously, to arbitrary accuracy, because a measurement carried out by a macro-apparatus to determine one of the canonical variables automatically interferes with the knowable values of the associated conjugate variable of that atom. In the thought experiment proposed, then, the canonically conjugate variables of each atom are the spin, S, and its orientation, Φ. For each of the hydrogen atoms, the uncertainty relations for these variables are:

$$\Delta S_1 \, \Delta\Phi_1 \geq h/4\pi \quad \Delta S_2 \, \Delta\Phi_2 \geq h/4\pi$$

As we have seen earlier, these inequalities are interpreted to mean that if one should measure the spin S_1 of atom 1 (or S_2 of atom 2) arbitrarily precisely, so that ΔS_1 (or ΔS_2) would be close to zero, then the measurement of this property of the atom must interfere with the precision with which the spin orientation could be specified, so that $\Delta\Phi_1$ (or $\Delta\Phi_2$) are correspondingly close to infinity, in accordance with the ratio $h/\Delta S_1$ (or $h/\Delta S_2$).

Nevertheless, in the situation posed by Einstein, Podolsky and Rosen, a correlation must persist between the spins of the two hydrogen atoms, as well as a correlation between the spin orientations of the two atoms, even though they are very far removed from each other, and therefore could be considered as noninteracting. It then follows from this correlation that if one should measure the spin of the first atom with infinitely sharp accuracy (so that $\Delta S_1 = 0$), the accuracy in knowledge of the spin orientation of that atom would be infinity, ($\Delta\Phi_1 = \infty$) but the correlation with atom 2 would mean that the spin of that atom would be as accurately known as that of atom 1; that is, $\Delta S_2 = 0$. That is to say, if one knows with certainty, from a measurement, that the spin of atom 1 is 'up', then the correlation with atom 2 would reveal that its spin would be 'down'—information deduced without in any way disturbing atom 2, because no measurement was made on this atom's physical properties. In a second experiment, one may be able to ascertain that its orientation has a definite value ($\Delta\Phi_1 = 0$) and this information would automatically reveal that the corresponding uncertainty in atom 2 is also zero, $\Delta\Phi_2 = 0$, without making any measurement on atom 2.

The point of this discussion is that the information about atom 2 must exist, independent of any measurement on it by a macro-apparatus. Thus, EPR argued that the atom of matter must necessarily have a complete description, independent of any measurement that may or may not interfere with it, in contradiction with the claim of the Copenhagen school. Thus it follows that all of the atoms of a material system have a complete description, independent of measurements. On the other hand, the quantum mechanical relations that lead to a comparison with the measured properties of an element of matter has such a form as to entail a 'weighting' of a particular linear operator (that is supposed to represent the act of measurement). The latter 'weighting function', which follows from *the fundamental law* of this elementary matter (quantum mechanics), has the form of a particular sort of *probability calculus*. Since the fundamental theoretical expression for the measured physical property of a single atom of matter depends on a probability function in quantum mechanics, this description is intrinsically incomplete, as far as the precise dynamical variables of the atom are concerned.

Einstein, Podolsky, and Rosen then concluded that so long as one should insist on interpreting the wave function of quantum mechanics in terms of *a single atom* (or a single elementary particle) then their analysis leads to a situation that is paradoxical: that the theory is both complete, because it is fundamental, and incomplete. This situation is called the 'Einstein-Podolsky-Rosen paradox'. Their final conclusion then implied that *at best,* quantum mechanics is an incomplete theory of elementary matter.

Of course, a logical paradox is unacceptable in any scientific theory: an alleged scientific theory lapses into a state of nonsense if indeed it entails a logical paradox. Einstein then concluded that so long as the premisses adhered to by the followers of the Copenhagen school were maintained, then their theory must be false. Nevertheless, one might get out of this trouble by reinterpreting the variables of quantum mechanics, particularly by abandoning the premiss that the wave function in Schrödinger's wave mechanics (or the state function in Heisenberg's matrix mechanics) refers to a single constituent of the entire system, relating *subjectively* to the measurements of the properties of this element alone.

If one should view the probability calculus in quantum mechanics in the same way as one interprets the probability calculus of statistical mechanics, where its solutions play the role of a weighting function for an entire ensemble of a very large number of constituents, providing a means of averaging the physical properties of the entire system, then the paradox would automatically disappear. With the latter point of view, the role of quantum mechanics would be to provide such a weighting function, *in addition to* the complete dynamical description that *underlies* the statistics. In this view, the incompleteness discussed by EPR would be natural, since one would still need the dynamical representation of the entire ensemble to complete it. Quantum mechanics alone would, in this case, not be intended to replace the underlying deterministic theory of matter in terms of discrete trajectories of the individual constituents of the entire system. Rather, it would simply be additional information that could be useful in estimating the averaged properties of the (assumed) ensemble of a very large number of atomic constituents.

This interpretation of the wave function of quantum mechanics, analogous to Boltzmann's distribution function in statistical mechanics (discussed in chapter two), is still deterministic since the underlying trajectories of the constituent elements of the entire system are there, even though our measuring instruments cannot specify all of them with equal precision at a given time. Neither would this theory be subjective in character compared with the Copenhagen approach because there would be no claim here that the physical properties of matter are in any way dependent on the manner in which they are observed. Of course, what is 'seen' is dependent on the way in which it is observed. But what is 'seen' is not all that there is. Further experimentation must then be used in order to exhaust all possible information of the system, which in principle would be a complete set of information.

Summing up, it appears that the EPR paradox was indeed a bona fide *logical* refutation of the Copenhagen interpretation of quantum mechanics. The paradox could, however, be removed within the particle model of matter if we should reinterpret the wave function of quantum mechanics as relating to a statistical weighting function for the entire system of atoms (or elementary particles) rather than taking it to relate to a single constituent atom.

It should be emphasized at this point that the latter interpretation of quantum mechanics, in terms of a statistical theory of an entire ensemble of atoms, is not the only possible interpretation of this theory. An alternative interpretation is the one in terms of a continuum field theory of matter, in line with the postulates of the theory of relativity, discussed in the preceding paragraphs. But investigating an interpretation in terms of an ensemble of (objective) elements of matter is still assuming that individual atoms exist which *underlie* the statistical description of quantum mechanics. Thus, it has been proposed by those investigators who wish to maintain the deterministic view of a system of real atoms that the incompleteness in the quantum mechanical (statistical) description alone should be complemented by introducing extra parameters upon which the dependent variables (wave functions) must depend. In these studies, it was also tacitly assumed that any actual observable does not relate directly to these extra independent variables. Thus they are called 'hidden variables'. They are there to complete the description of the (now assumed) predetermined trajectories of the constituent elements of matter of macroscopic proportions. Thus, the attempt of the 'hidden variable' approach to quantum mechanics is to remove the subjective aspects from the fundamental theory of matter, while still maintaining the outward probability calculus.

The latter view of atomic matter adopts the philosophic standpoint of *realism,* rather than the positivistic standpoint of the Copenhagen school. Nevertheless, it is a position that Bertrand Russell has called 'naive realism', assuming that the (actually unobservable) atoms of matter are the way they are supposed to be because of the way that we observe macroscopic matter to be. In contrast, the theory of relativity as a theory of matter takes the stand of 'abstract realism'. It is the latter epistemology that Einstein adhered to, after his research evolved to the theory of general relativity and the continuous field approach. Still, because the arguments of the EPR paradox indicated that a hidden variable approach *may* resolve the problem, most physicists have assumed that it was Einstein's contention that one must proceed to a hidden variable theory to resolve the problem of quantum mechanics.

This assumption is not at all true. Einstein's actual choice of a theory to resolve the problem was one that is based fully on his theory of general relativity as a theory of matter, which rejects any sort of atomistic model, replacing it with the continuous field concept as fundamental. (See Einstein's comment to Bohm in the last paragraph of Note 65.)

Bohr's Reply to Einstein, Podolsky, and Rosen

Bohr replied to Einstein, Podolsky, and Rosen by saying that their conclusion about the incompleteness of quantum mechanics was fallacious because it was *out of context*.[64] Bohr's interpretation of quantum mechanics is that it is a set of rules regarding the outcomes of measurements by macroapparatuses on the physical properties of micromatter—*when the measurements are carried out*. In this view, quantum mechanics does not deal with the history of the elements of matter, from an earlier time, say when two particles might have been bound, to a later time, when they would be free of each other. When one observes the atoms in a bound state, as in observing the properties of H_2, or when the separated, unbound atoms are observed on their own, these would refer to two separate sorts of measurements; they must therefore be represented by different sorts of wave functions. Thus Bohr rejected the EPR claim that quantum mechanics is incomplete, claiming that its opponents were not interpreting the elements of the theory correctly. It was Bohr's contention that, based on the axiomatic starting point of quantum mechanics, it is as complete a theory of elementary matter as it possibly can be.

At this stage of the debate between Bohr and Einstein (around 1935), Einstein realized that its character had changed from questions about physics to questions about epistemology. Einstein did not agree with Bohr and Heisenberg that a probability calculus *necessarily* expresses a maximally complete knowledge about elementary matter. On the other hand, this was Einstein's opinion, because of his epistemological position of abstract realism, in contrast with the Copenhagen school's logical positivism. At that time, Bohr's starting assumptions did successfully lead to nonrelativistic quantum mechanics and its predictions that agreed very well with low energy atomic physics experimentation. Further, Einstein did not provide an alternative theory that might be equally successful in predicting these same data. Still, it is true that to achieve agreement with the data is a *necessary* condition imposed on a scientific theory, *but it is not sufficient.* For a theory must also be logically and mathematically consistent, predicting unique results for unique physical situations.

To be acceptable, Einstein also believed that a theory must be *simple.* By this he meant that the theory should be *whole* and *complete.* He did not see that quantum mechanics met this standard. Further, when the attempt was made to fuse quantum mechanics with the theory of relativity, insurmountable difficulties arose, as we will discuss in detail in chapter ten. We must then ask what the theory of relativity is all about when viewed as a fundamental theory of matter. Before going on to this question in the next chapter, the next section of this chapter will be devoted to a more detailed outline of the current attempts to restore determinism to quantum mechanics by incorporating the hidden variable approach.

The Hidden Variable Approach

The activity in hidden variable theories in the early 1950s was motivated in part by the attempt to resolve the Einstein-Podolsky-Rosen paradox. This work was largely started by David Bohm and his co-workers,[65] though it was studied much earlier from a slightly different point of view by de Broglie (discussed in chapter five).

De Broglie did not give up the view of particles as real, objective entities that move along predetermined trajectories, independent of whether or not measurements may be carried out. But he also wanted to incorporate into his theory the empirically verified wave nature of the particles, as described with Schrödinger's wave equation. De Broglie was then led to suggest that there must be a second solution of the 'completed theory' that relates directly to the trajectories of the singular particles of matter that are there, whether or not they are seen. The latter were then the 'hidden variables' of his second solution resolution of the problem of quantum mechanics, an attempted resolution that restores determinism by going to a view of predetermined trajectories that, nevertheless, are hidden from direct observation.

The idea of a hidden variable approach was also indicated in order to meet Schrödinger's objection to the single particle interpretation of the quantum mechanical wave function: while one might associate a particle of matter with a highly localized wave that has a definite wavelength, this particle wave must eventually *disperse* into a packet of waves, each having a slightly different de Broglie wavelength. Thus, one may initially identify a material particle with a constant wavelength according to de Broglie's formula, $p = h/\lambda_o$, but in due course this particle would become many particles simply by virtue of the dispersion of the original wave into many waves, with different wavelengths, $\lambda_1, \lambda_2, \ldots$

Schrödinger then asked: how can one associate the original particle-wave with a single particle in the first place, if its history cannot be followed, maintaining its integrity as a particle?[66]

The dispersion that Schrödinger refers to occurs simply by virtue of the interaction of the single original wave with any other material object, such as any sort of measuring apparatus. He then asserted that it would be impossible to objectively identify a single wave with a single particle because a single particle would endure while a single wave would not. He was then led to the same conclusion that was arrived at in the EPR analysis: that the quantum mechanical wave function must relate to an entire ensemble of particles of matter, rather than to a single particle, if there is to be logical consistency in this model.

Nevertheless, if one insists on identifying the quantum mechanical wave function with a single particle, a resolution of Schrödinger's problem may be achieved by arriving at a *dispersion-free* wave for the particle. Can this be done? It was de Broglie's idea that it was indeed the function of his second (singular) solution to hold the original undispersed wave together—yielding the undispersed state. He also believed that the nonlinear feature of his added equation, that in turn predicts his second solution, predicts that the particle trajectory is *buried inside of* the Schrödinger wave, preventing the disintegration of the original particle (i.e. its dispersion into a family of new waves). But, as we have discussed earlier, the full, formal expression of de Broglie's research program has not been completed to this time.

Summing up, de Broglie's form of a 'hidden variable' theory interprets the formal expression of quantum mechanics as a set of rules for determining the statistical distribution functions for an ensemble of particles, but the 'hidden parameters' are to relate to a 'predetermined' set of trajectories that describe the objective particles of matter that are there all the while. It is the latter set of variables that do not disperse, and thus that may be identified with the localized particles that do indeed endure.

Von Neumann's Proof

There is a well known proof, ascribed to John Von Neumann, showing that there cannot be any dispersion-free states within the formal structure of quantum mechanics.[67] At first glance, this proof seems to rule out the possibility of any hidden variable theory. Nevertheless, it is important to note that Von Neumann's proof relies on the validity of the *principle of linear superposition* as well as the general description in terms of 'Hilbert space'. It is then possible that when de Broglie's nonlinear addition to the linear wave equation is expressed rigorously, Von Neumann's proof would not apply to this 'hidden variable' theory.

Von Neumann's proof has to do with the following: If $\hat{A}$, $\hat{B}$ and $\hat{C}$ are any three 'measurement operators', as we have discussed previously, such that one of them may be expressed as a linear sum of the other two

$$\hat{C} = \alpha\hat{A} + \beta\hat{B}$$

then the expectation values of these operators (i.e. their averaged measured values) are also linearly related:

$$<\hat{C}> = \alpha<\hat{A}> + \beta<\hat{B}>$$

For the assumed 'dispersion-free' states, there is no distinction between the expectation values of the operators above and their discrete eigenvalues, say A_n, when ψ_n is a dispersion-free state function, i.e.

$$\hat{A}\psi_n = A_n\psi_n \textit{ implies } <\hat{A}>_n = A_n$$

On the other hand, the eigenvalues, B_n and A_n, are not generally additive. That is, it is generally true that

$$C_n \neq A_n + B_n$$

They do add *only* if the corresponding linear operators, $\hat{A}$, $\hat{B}$ *and* $\hat{C}$, commute with each other, i.e. $\hat{A}\hat{B} - \hat{B}\hat{A} = 0$.

An important example to demonstrate this is the way in which an electron couples to an external magnetic field. If we define the orientation of this field as the *z*-direction, then it turns out that the operator that describes this coupling depends on the following matrix:

$$\sigma_z = \begin{pmatrix} 1 & 0 \\ 0 & -1 \end{pmatrix}$$

The $+1$ in this matrix denotes that the 'spin' of the electron is parallel to the external magnetic field and the -1 denotes that it is antiparallel. (This is called the 'Pauli spin operator'). The eigenvalues of this operator are then the two possible values, $+1$ and -1. These may be identified with the corresponding states, ψ_1 and ψ_{-1}. They are the states associated with a measurement of the electron's spin (magnetic moment), when it would be either parallel or antiparallel to the externally applied magnetic field.

Suppose now that we should measure the *x*-component of the electron's magnetic moment. This measurement operator would depend on the 'Pauli matrix',

$$\sigma_x = \begin{pmatrix} 0 & 1 \\ 1 & 0 \end{pmatrix}$$

If one should now 'operate on' one of the eigenstates of σ_z with σ_x, say the state ψ_1, this state would be found to 'disperse' into a different state of σ_z, particularly, ψ_{-1}. We may denote these states as:

$$\psi_1 = \begin{pmatrix} 1 \\ 0 \end{pmatrix} \text{ and } \psi_{-1} = \begin{pmatrix} 0 \\ -1 \end{pmatrix} \Rightarrow \sigma_z \psi_1 = \psi_1, \sigma_z \psi_{-1} = -\psi_{-1}$$

However, according to the rules of matrix multiplication,

$$\sigma_x\psi_1 = \psi_{-1} \quad \text{and} \quad \sigma_x\psi_{-1} = -\psi_1,$$
$$\Rightarrow \sigma_x\psi_1 \not\propto \psi_1, \text{ etc.}$$

Thus, we see that the eigenvalues of the operators that refer to measurement of the electron's spin in the respective z-direction, parallel to the applied magnetic field, and the x-direction, perpendicular to the magnetic field, are *not additive.* As we have discussed previously, this is interpreted to mean that the z- and x-components of the electron's magnetic moment (which depends directly on its 'spin') are not determinable simultaneously. Mathematically, this idea is expressed in terms of the noncommutation of the corresponding spin operators when they 'operate on' one of the eigenstates of σ_z or the other. The states of this system are then not dispersion-free, because

$$(\sigma_x \sigma_z - \sigma_z \sigma_x)\psi_1 \neq 0$$

The hidden variable theory is the attempt to generalize the form of quantum states in such a way that they do become dispersion-free. This is done by generalizing the space-time coordinate system, in which the state functions are mapped, to an expanded parameter space:

$$\psi(\boldsymbol{r}, t) \rightarrow \psi(\boldsymbol{r}, t, \lambda)$$

where λ indicates the 'hidden parameters'. Their change with respect to the time measure, t, then denotes the actual velocity of the singular trajectory of matter. That is, it is next assumed that the state functions depend on $\boldsymbol{r}$ and $\lambda(t)$, which denotes the 'deterministic' trajectory of the particle; the state function then is viewed as the following dependent variable, $\psi(\boldsymbol{r}, \lambda(t))$.

Finally, proceeding from the single particle to a many-particle system, the state function becomes dependent on the three coordinates of ordinary space, $\boldsymbol{r}$, the time parameter, t, and (for an n-body system), n hidden parameters, which are functions of the single time parameter,

$$\psi[r, \lambda(t)] \rightarrow \psi[r, \lambda_1(t), \lambda_2(t), \ldots, \lambda_n(t)]$$

The current density that relates to the velocity of any one of these constituent trajectories, $\lambda_m(t)$, then depends on the rate of change of the *dependent variable* (the state function ψ) with respect to the hidden variable that is the *independent variable,* $\lambda_m(t)$, to represent the statistics of the many-body system, whose determinism is in the full set $\{\lambda(t)\}$.

This is a rough sketch of the role of variables in the hidden variable approach to quantum mechanics. It must be investigated further to see if indeed it could resolve the problem of micromatter in a logically consistent manner, maintaining the statistics of quantum mechanics while also maintaining an underlying deterministic (though 'hidden') description of the constituents of the system, that are its particles. It would be pertinent if this approach could resolve the EPR paradox, as well as satisfying Schrödinger's requirement of formulating a dispersion-free wave theory of elementary matter.[68]

At the present stage of this research program, it doesn't seem to be working. One of the troubles is that the many-body wave function in this formulation still depends on all of the hidden variables, $[\lambda_1, \lambda_2, \ldots, \lambda_n]$ for any given trajectory of the system, $\lambda_i(t)$. This means that the fundamental description of a single particle of the system, say any given electron, depends on the trajectories of all of the other particles of the system; it has no individual description that is *local*. Such *non-locality* in the description of matter is also a feature of a continuum field theory of matter. But the latter rejects the particle concept altogether. That is to say, the hidden variable description of the elementary particles of matter, as it now stands, is a theory of a given particle in terms of a wave function that is supposed to accompany the actual trajectory of this particle, that still depends on the variables of all of the other particles of an entire ensemble. Thus, it does not yet have the form of a 'particle theory', because it is *nonlocal:* the trajectory of a particle depends here on the trajectories of other particles in other places.

The question then arises: can there be a hidden variable theory that is local? There have been several analyses in the past that indicate that such a scheme of locality within the structure of quantum mechanics—using the required 'Hilbert space' structure for the functions that are associated with the probability calculus of quantum mechanics—cannot in fact exist. A possible resolution might then be to abandon the particle aspect altogether, reducing the 'wave-particle dualism' in its philosophical basis to a 'wave monism'. This would then imply a radical change in the formal expression of the theory of elementary matter. Nevertheless, in view of the empirical successes of *nonrelativistic* quantum mechanics, one must still maintain the formal structure of the (Schrödinger or Heisenberg) representation of quantum mechanics as a *mathematical approximation*.

This may be an approximation for a different theory altogether. The conceptual approach of quantum mechanics, strictly, would be false if either the hidden variable approach or Einstein's continuum field approach is true. For example, as we will discuss in more detail in chapter ten, Einstein's view of matter in general relativity is based on the notions of full objectivity of the laws of nature, continuity, nonlinearity and determinism. These ideas contrast with the fundamental ideas of quantum mechanics, of irreducible subjectivity, atomism, linearity and nondeterminism. Einstein saw quantum mechanics, though empirically valid in a particular mathematical approximation, as truly following from a theory of matter that is not based on any of the ideas that were propounded by Bohr and Heisenberg—the so-called Copenhagen school. The question then followed: is it possible that indeed the basis of the theory of relativity could also lead to the same successful results that came from quantum mechanics, but from its (entirely different) axiomatic basis? We will see later on that this is in fact a distinct possibility, with a proper generalization imposed.

It was the *principle of relativity,* expressed by Einstein in the early decades of the twentieth century, that formed the axiomatic basis of his theory of relativity, the other revolution of contemporary physics. In the following chapter we will outline its main philosophy. We will then outline attempts that have been made to fuse these ideas and their mathematical expression with the revolution of the quantum theory of measurement. The fact that these two major developments in contemporary physics have not yet been successfully unified constitutes one of the major crises in modern science. For in their own domains of explanation, each of these theories 'worked' well, so long as it could be assumed that they did not overlap. But it soon became evident that these theories could not be treated separately from each other in a logically consistent way. It turned out that if one of these theories was to be accepted, the conceptual notions of the other would have to be logically excluded. This fascinating development in physics, because of the simultaneous appearance of two revolutions that in principle logically exclude each other, then implied further adventure into unknown territory, to develop the ideas that will likely lead to a third revolution in physics, probably to come to its full fruition in the twenty-first century. Whatever form it will have, it will likely retain features of both the theory of relativity and quantum mechanics. But which of these will have more influence on the future of physics?

Before the reader comes to a conclusion on this, let us discuss in more detail what the theory of relativity is really all about and how its basis compares with that of quantum mechanics; we shall consider the two as opposing theories of elementary matter.

Eight
The Theory of Relativity What's It All About?

The Principle of Relativity

The theory of relativity is primarily based on a single idea: *the principle of relativity*.[69] This is an assertion that the laws of nature must be independent of the reference frame in which they are expressed. This idea is equivalent to saying that the laws of nature must be fully *objective*. This principle in physics then seems to be on a different plane than the other physical principles, since it appears at first glance to be tautological, like saying that 3 = 3. This is so because a 'law' wouldn't be a 'law' in the first place if it were not totally objective. For example, a 'traffic law' asserts that whenever one comes to a red light, one must stop, whether in New York, Tokyo or Jerusalem. It is a universal statement about the rules of traffic. To say that a law is universal is then simply saying something about the definition of 'law'. Nevertheless, the principle of relativity is not really tautological because it entails two tacit assumptions that are not necessarily true, that is, two assumptions that are indeed contingent on nature.

The first of these contingent assumptions is that there are laws of nature in the first place, laws that underlie the outward manifestations of matter. The second contingent assumption is that these laws may be comprehended and expressed with a language that is sufficiently rich that we would be able to make precise predictions and test them with accuracy.

The first of these contingent assumptions implies that the universe is indeed characterized by an *underlying order*—in any of its domains, from *fermis* (and smaller) to *light-years* (and

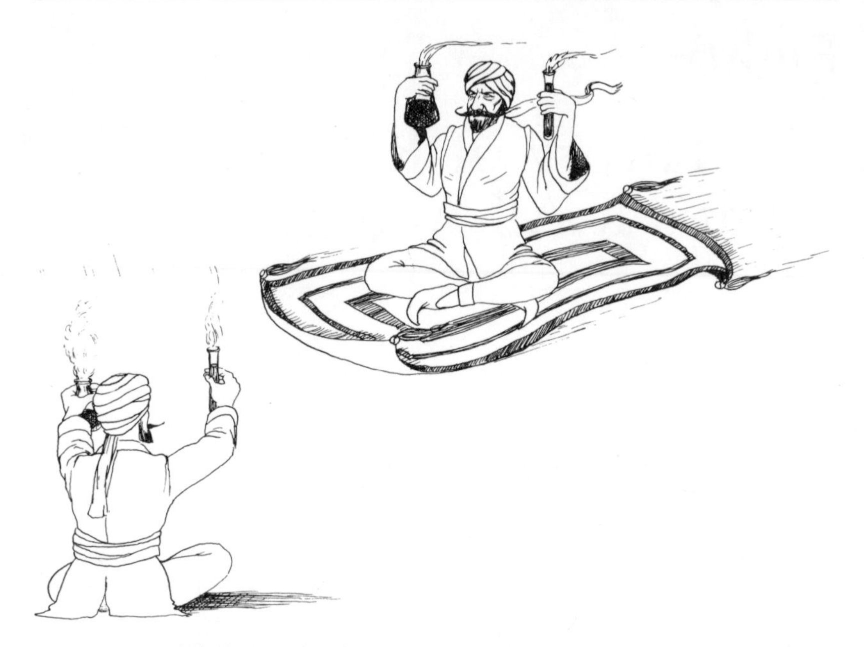

Figure 8.1. The Principle of Relativity. *A stationary scientist and one who moves on a flying carpet perform similar experiments to deduce a particular law of nature. If each watches the other's experiment and compares it with his own, the principle of relativity requires that the compared expressions for this law must be in exact correspondence.*

greater). When it is expressed in terms of cause and effect, the order that is referred to then signifies that for every effect of the universe (from the smallest to the largest) there must be a logically connected cause. It follows that, in principle, such a view rejects any acausal theory of matter, in any domain and in any degree. For example, it would be in conflict with the claim of a fundamental acausality in the measurement process in microphysics, according to the Copenhagen interpretation of atomic matter, discussed in the preceding chapter.

The second tacit assumption, implicit in the principle of relativity, is that it should be possible for us to find a language that would express the laws of matter in a precise way. Recall that it was Galileo who argued that this must indeed be the case, and that "in order to read the great book of nature, one must learn the language of Mathematics". That is to say, from the point of view of physics, mathematics serves as a *tool:* it plays the role of a language whose only purpose in this context is to facilitate an expression of the laws of nature.[70]

There have been stages in the history of physics when the discovery of particular mathematical forms to express physical laws (such as Newton's discovery of the calculus to represent variable motion) was so successful that scientists tended to equate mathematics with science. In my view, the latter was indeed a very unfortunate development since it had the effect of stagnating progress because many ceased to question the physical hypotheses and instead concentrated on the mathematical expressions of a *given* hypothesis. Mathematics is a type of art form. It is based on a beautiful network of structures that are the invented axioms and theorems, and their implications (as in Euclidean geometry). But there is no reason to believe that any of these axioms or theorems may in any way be related to nature. Mathematics is analytic knowledge. On the other hand, the starting axioms in science, and physics in particular, are contingent on nature. Their truth values may only be tested by comparing the conclusions of these scientific hypotheses with the experimental facts. And then, agreement may be achieved at one stage of physics, but when better resolution is obtained with experimental devices and new sorts of experiments are created, or a broader theoretical view is considered anew, one must be ready to abandon the 'truth' of the earlier scientific hypotheses if they do not hold up in the light of the new observations and theoretical analyses. Indeed, this is the way in which we achieve progress in our pursuit of *scientific truth*. Thus, 'mathematial knowledge', such as: 'within a particular arithmetical logic, 2 + 2 = 4', *is necessarily true.* But 'scientific knowledge' is not necessarily true, at any stage, even though we continually seek common threads of ideas that may persist throughout the different periods of the history of science.

In regard to the mathematical language of physics, we have found that its most precise form (thus far) entails a set of relations that define a continuous change of *dependent variables* with respect to (at least) *four independent variables*. Three of the independent variables are called 'spatial parameters', labelled x_1, x_2, x_3; the fourth is the temporal parameter, labelled t. The independent variables are called by these names because we may correlate them with measures of spatial extension and temporal duration, respectively. However, it is important to note that these four variable parameters are not identical with 'space' and 'time'; they are only language elements that are invented for the purpose of expressing *measures* of space and time (perhaps for want of a better language). It is especially important to take note of this difference between space and time and the words that are used to express their measures in the theory of relativity. This will become clear later on in this chapter and especially in the following chapter, where we will see how not paying attention to this difference leads to real paradoxes. These paradoxes are the sources of some very serious conflicts in contemporary relativity physics, for example in regard to what has been called the 'twin paradox', to be discussed in chapter nine.

Herein lies a very important difference between the theory of relativity, as a fundamental theory of matter, and the classical theories that compete with it. In the theory of relativity, the spatial and temporal parameters do not, in themselves, denote any sort of reality, while they do so in the classical atomistic theories, such as the Newtonian view of matter or even the quantum theory. In these latter approaches, space and time relate to an absolute quality of matter, called 'trajectory'. In contrast to this, relativity theory derives its name from the assumption that the language of space and time variables is relative to the reference frame that is used in order to express *absolute* laws of nature. Thus, the theory of relativity does *not* assert that 'everything is relative', as many popularized accounts have falsely claimed. This would be the view of the philosophy of relativism. In contrast, the theory of relativity asserts that there are some things about the physical world that are absolute. One of these is the universe itself. The universe cannot be any sort of relative entity because it is, *by definition,* all that there is. Thus there cannot be anything it is relative to. Now if the universe is characterized by its laws of natural phenomena, then these must also be absolute: that is, the laws of natural phenomena must have forms that are *totally objective*. This means that they must be in one-to-one correspondence in all possible frames of reference. But this conclusion, following from the absoluteness of the universe as a whole, is nothing more than the statement of the *principle of relativity*. It is then the job of the theoretical physicist to find the most efficient language for the expression of the frame-independent laws of natural phenomena.

The 'relativity' of the language of space and time measures, to express absolute laws of nature, may be understood with the following analogue: Before Newton, an Englishman may have asserted the following law of nature: 'Whatever goes up *must* come down'. At the same time, a Frenchman may assert: 'Ce que s'élève, *doit* descendre'. Each of these sentences says the same thing. If the Englishman did not understand French and if the Frenchman did not understand English, they would not know that each was saying a sentence with the *same meaning*. While the languages are relative to the reference frames of England and France, the meaning expressed in each of the sentences is in one-to-one correspondence with the other—the meaning is absolute. If the Englishman should learn French and discover that the Frenchman was saying something different than what he was saying, then according to the principle of relativity he would have to doubt the truth of either his statement or that of the Frenchman. For if a statement about physical phenomena is scientifically true, according to Einstein's principle of relativity it must correspond exactly in all possible reference frames. This is the main idea of Einstein's theory of relativity.

The Transformations of Space-Time Languages

Albert Einstein first arrived at his 'principle of relativity' when he recognized that at least one of the laws of nature (the law of electromagnetism, expressed in terms of Maxwell's equations) does not maintain its form in all possible relatively moving frames of reference, unless the classical idea of the relativity of space is extended to the relativity of a fused space and time, called 'space-time', in the natural expression of that law. What this means is that a purely spatial measure, expressed in one reference frame, must be represented in other relatively moving reference frames in which one may wish to express the same law of nature, in terms of mixtures of spatial and temporal measures that are prescribed by the unique space-time transformations. The uniqueness of these transformations must be maintained, independent of any particular law of nature, in the same sense that there must be a unique translation between English and French, independent of any particular sentence that one wishes to translate, or any particular subject.

The fusion of 'space' and 'time' into 'space-time' is nothing more than an enrichment of the language that is necessary in order to express all of the laws of nature, objectively. It must be noted that this is *not* a fusion of physically different entities, such as a fusion of physical extension and physical duration. If the latter would be true, then it may rightfully be asked: if we can travel in space, why would it not be possible to also travel in time, forward or backward? The answer is that travel in space and travel in time (if it is possible) has to do with a sequence of experiences. We can have these experiences in traveling from New York to Miami. But these experiences are not space; they are, rather, our moving our bodies (on a train, an airplane, a bicycle, and so forth) from one physical point to another. The space we talk about in relativity theory is a *measure*. We invent this measure in order to represent spatial extent, just as we invent the twelve numbers on the face of a clock in order to represent the duration of the unwinding spring behind the face of the clock. But the invented number system that we use to measure the time sequence is not in itself 'time'. Similarly, the space-time measure is not a fusion of physical space and physical time; it is more like the fusion of Latin, Saxon, French, and Gaelic to form the English language.

What Einstein saw in the early stages of relativity theory was that the form of Maxwell's equations—the law of electromagnetism—is in one-to-one correspondence in all possible reference frames that move in a straight line relative to each other, at constant relative speeds, under the same set of transformations that leave the following combination of four squares unchanged:

$$s^2 = [(ct)^2 - (x_1^2 + x_2^2 + x_3^2)] \quad (1)$$

where c is the speed of light in a vacuum, t is a time measure and $(x_1^2 + x_2^2 + x_3^2) = r^2$ is the (square of a) three-dimensional space measure. Such reference frames are called 'inertial' and the invariance of the laws of nature under these transformations is the requirement of the *theory of special relativity,* where the adjective 'special' refers to the special type of relative motion that is called 'inertial' (that is, constant relative speed in a straight line). When the motion between the reference frames in which we wish to compare laws of nature is not inertial, the theory refers to 'general relativity'.

Einstein recognized that the invariance of the interval s^2, in defining the transformations that leave the laws of nature unchanged (objective) in different inertial frames of reference, requires that the speed of light, c, must be a universal constant: that is, the same as measured in any frame of reference, from any particular observer's view. As we will discuss below, this universal constant first appeared as a conversion factor, to express relative temporal measures in units of spatial length. However, identifying this constant with the predictions of Maxwell's equations of electromagnetism, it turns out, numerically, to be the speed of light.

Einstein then proposed that the space-time transformations that preserve the objectivity of Maxwell's equations for the electromagnetic laws are the same transformations that preserve the forms of all of the laws of nature. In the Maxwell theory, the speed of light turns out to be the speed of propagation of the electromagnetic force between interacting, electrically charged bodies. Einstein's generalization then implied that the speed of light, c, is indeed the maximum speed of propagation of *any* force between interacting matter. His idea, then, replaced the classical Newtonian notion of action-at-a-distance with the notion that all forces (including the gravitational force) must propagate at a finite speed.

The Light Cone

Just as the constancy of the (squared) radius, $r^2 = x_1^2 + x_2^2$, in 2 dimensions, with respect to rotations, defines a *circle,* and the constancy of the (squared) radius $r^2 = x_1^2 + x_2^2 + x_3^2$ with respect to rotations in a three-dimensional space defines a sphere, so the invariance of the four-dimensional interval s^2, with one squared coordinate positive and the other three negative (equation (1)) defines a *cone,* geometrically. When the magnitude of this interval is zero, $s = 0$, the invariance of s to transformations between different inertial frames implies that

$$ct = r \rightarrow ct' = r'$$

where $r = (x_1^2 + x_2^2 + x_3^2)^{1/2}$ is the three-dimensional spatial interval and $c = c'$ denotes the universality of the speed of light, as discussed above (it will be logically derived in the next section).

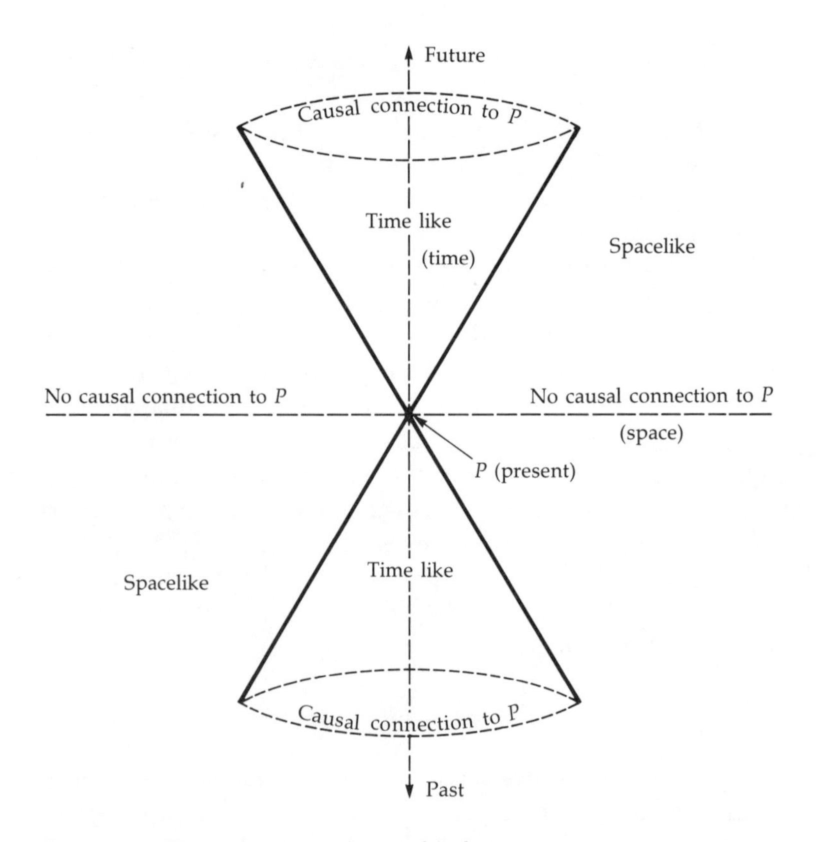

Figure 8.2. The Light Cone of Special Relativity.

Thus we see that in the special case where $s = 0$, the spatial and temporal intervals connect the vertex of the 'light cone'—the 'present'—with the other points on the surface of the light cone (see figure 8.2). The meaning of these intervals is that they are a representation of the space-time language relations for the propagating force itself—such as a light beam, propagating the electromagnetic force from one bit of charged matter, at one place and time, to another bit of charged matter, at another place and at a different time.

The space-time intervals, s, that are not on the light cone relate to the interacting matter itself, at different space-time points. When the magnitude of s is positive, then

$$ct > r$$

where r is the spatial separation of the interacting matter and t is the time it takes the force exerted by one bit of matter to get to the other. These intervals, $s > 0$, correspond to the space-time points at the vertex of the cone (the 'present') connecting to the other points that are *inside of* the light cone. Such intervals are called 'timelike'. They relate to causally connected coordinates—the matter components separated by such intervals interact by means of a force that propagates between them in a time t, which is enough time to enhance the cause-effect relation when these bodies are separated by $r < ct$.

When the magnitude of s is negative, then

$$ct < r$$

Geometrically, these correspond to intervals that connect the vertex of the light cone to all of the points *outside of* it. These intervals, $s < 0$, are called 'spacelike'. They refer to spatial separations between inertial matter that is too great for a signal that travels between them at the speed c to arrive in the time interval t. Thus, the matter at 'spacelike' separations is not causally connected.

We see, then, that only the portion of the space-time language that may be geometrically related to intervals between the present (the vertex of the light cone) and the points inside or on the light cone may be used to represent interacting matter (that is, with causal connection) in relativity theory. In contrast, the entire space and time is used in the language of the Newtonian theory of matter. This is because the classical theory is based on the idea of *action-at-a-distance* which does not entail any time interval for the interactions between distant masses to occur, no matter how far apart they may be.

As we see in figure 8.2, there are two portions of the light cone: one in forward time and the other in backward time. With the use of the space-time language in the expression of the laws of nature, the forward part of the light cone represents 'predictive' causal connection, with the temporal interval, $+t$, denoting the time it takes for the cause (at the present) to give rise to an effect, at (r,t). Similarly, the points in the backward light cone, $-t$, denote the past, and express 'retrodictive' causal connection. In this context, 'future' and 'past' are only meaningful in terms of the causal connection in laws of nature, expressing cause-effect relations. The intervals that connect the points in the backward light cone to the present simply relate to causes in the past for effects in the present, as expressed explicitly with the laws of nature that use this space-time language. That is, the backward light cone does not relate to any physically connected relation which implies that an effect precedes a cause.

The 'future' and the 'past' are only meaningful in the context of the theory of relativity in terms of causal connection, as indicated laws of cause-effect interactions. Thus, the points of space-time outside of the light cone, that is, the points connecting the 'present' with spacelike intervals, may not be associated with any time order, toward the past or toward the future.

Note also that the invariant form of s^2, according to equation (1), is more symmetric than it has to be. For according to relativity theory itself, the transformations between reference frames have to do with relative *motion,* which in turn is only a *continuous* type of change of the space and time coordinates. However, s^2, as given in equation (1), is also unchanged under the discrete reflections in space and time:

$$t \rightarrow -t \text{ (time reversal) and } \mathbf{r} \rightarrow -\mathbf{r} \text{ (space reflection)}$$

The removal of the time and space reflections from the total set of transformations of (special and general) relativity theory has been found to have a profound effect on the formulation of the most general expressions of the laws of nature.

The Universality of the Speed of Light

Why is the speed of light asserted to be a universal constant in the theory of special relativity? The answer may be found in the conceptual role that the space and time parameters play as the elements ('words') of a language whose only purpose is to facilitate an expression of objective laws of nature. As we have discussed in the preceding section, if a spatial measure as expressed in some frame of reference (called K') is r', then it must generally be expressed as a linear combination of a spatial measure, r, and a temporal measure, x_o, in a different reference frame, K, that moves relative to the K' frame, i.e.

$$r' = a_1 r + a_o(x_o) \quad (2)$$

The coefficients (a_1, a_o) of this transformation (translation of languages) are dimensionless and, in special relativity theory, they are independent of the particular space-time points where they are used (though not so in general relativity theory). It then follows that the measures of space and time, r and x_o, must be expressed in the same units. If r is measured in spatial units, say centimeters, then the time measure x_o must also be expressed in centimeters, even though it refers to a measure of time, that is, seconds. It then follows that t seconds must be expressed in this theory as $x_o = ct$ centimeters, where c is a conversion factor with dimension of centimeters/second. That is, the conversion factor, c, must have the dimension of a speed.[71] Thus, in one reference frame, K, the time measure would be expressesd as $x_o = ct$ cm, in a different inertial frame it would be expressed as $x_o' = ct'$ cm, in a third inertial frame it would be expressed as $x_o'' = ct''$ cm, and so on.

From the imposition of the principle of relativity, it can be seen that the objectivity of the laws of nature requires of the space-time language that a spatial measure in one reference frame must be expressed as a mixture of a spatial and temporal measure in other inertial frames. This *necessitates* the expression of the time measure in the different frames, $(t, t', t'', \ldots)$ **sec** as $(ct, ct', ct'', \ldots)$ **cm.** Further, if we are to sort out the time measure from the space measure in any particular space-time frame of reference, it is logically necessary that all reference frames must use the same conversion factor, c **cm/sec.** That is, the expression of the laws of nature in accordance with the principle of relativity logically requires that the conversion factor, c, must be a *universal constant.*

The expression of any law of nature in a way that would be consistent with the principle of relativity must then require the appearance of the conversion factor, c, whenever the time measure t appears in the corresponding equations. When the comparison of electromagnetic phenomena and the equations of the theory that is supposed to make these predictions is considered—that is when we apply these ideas to the Maxwell field equations for electromagnetism—we find that indeed the conversion factor, c, is numerically equal to the speed of light in a vacuum, which then corresponds to the speed of propagation of the electromagnetic force between charged interacting matter. Once this value of c is established, its constancy in the language of all of the laws of nature implies that this same constant will appear in all of the other laws of physical phenomena.

This means that there is nothing special about the laws of electromagnetism in particular, except that they happened to be the first to be investigated in the context of relativity theory—thus the universal speed c was called 'the speed of light'. If the laws of gravitation had been the first to be investigated in the context of relativity, c might have been called thenceforth 'the speed of gravity'. Generally, c is a universal constant (independent of any reference frame) that is the maximum speed of *any* type of force between interacting matter, according to Einstein's theory of special relativity.

The implication of the latter conclusion, that rules out action-at-a-distance, is also in conflict with some present day models in elementary particle physics. For this is a view that implies that the inertial particles that are supposed to mediate other (non-electromagnetic) interactions, such as the virtual pi mesons that mediate the nuclear interactions, which are inertial and therefore propagate at speeds less than c, must in themselves be composites of smaller particles, which in turn are mediated by forces that propagate between them at the speed *c*.

In his original writings on the theory of special relativity, Einstein asserted that this theory is based on *two* axioms: 1) the principle of relativity and 2) the universality of the speed of light. What we have now seen is that the first of these axioms logically implies the second. That is, while the universality of the speed of light is a feature of interacting matter, according to the theory of relativity, it is not an independent axiom of the theory. Thus, this reasoning gives us the stronger statement about relativity theory, that it is based on only one underlying axiom: the principle of relativity. Such a conclusion not only makes the theory stronger than it was previously thought to be, but it also satisfies Einstein's criterion about *his feeling* that a valid theory in nature must entail a maximum amount of conceptual simplicity. (This is sometimes referred to as his 'principle of simplicity'.)

Relation to Quantum Mechanics

From what has been said so far in regard to the logical basis of the theory of special relativity, it is clear that if the formal expression of quantum mechanics is indeed to represent a fundamental law of nature, then its equations (either in the form of Schrödinger's wave mechanics or Heisenberg's matrix mechanics, discussed in chapters five and six) must conform to the rules of space-time transformations of relativity theory, that would force these laws to be objective with respect to all possible inertial frames of reference. That is to say, the *form* of the equations of quantum mechanics, as a fundamental law of matter, must be preserved when transforming their expression to all possible reference frames—with the space-time transformations that leave the interval s^2 (in equation (1)) unchanged (except for the reflections, discussed above). Thus, all of the *probability statements* of the quantum theory must be in one-to-one correspondence in all possible inertial frames, according to Einstein's theory of special relativity.

This requirement entails the propagation of the interaction between matter at a finite speed *c*, whereas the equations of quantum mechanics entail instantaneous action-at-a-distance, according to the following scenario: a measurement is made on a property of micromatter by a macroapparatus; the measurement is represented in terms of the effect of a linear operator that acts on the state function associated with the measured micromatter. Then, *instantaneously*, the action of this apparatus causes the micromatter to respond with a 'proper' value for the measurement, as expressed in terms of the eigenvalue associated with that particular state function. But instantaneous events in one reference frame would not be described as instantaneous in other reference frames, according to relativity theory. Thus we see that the quantum mechanical form of the laws of micromatter would not be preserved in transforming them to other reference frames. This is in conflict with the requirement of the theory of relativity. We will discuss this problem in more detail in chapter ten.

From Special Relativity to General Relativity

Extending now to the theory of general relativity, it follows that all of the relations of the probability calculus, called 'quantum mechanics', must be objective (invariant in form) with respect to transformations between reference frames that are in artibrary sorts of relative motion. The trouble here is that the transformation coefficients in general relativity theory (such as the coefficients a_o, a_1 discussed above) are *nonlinear*—their values depend on the space-time points where they are used; on the other hand, the equations of quantum mechanics *are in principle linear,* in accordance with the underlying 'principle of linear superposition' of this theory which has to do with its interpretation as a particular sort of probability calculus. Generally, then, there appears to be a fundamental incompatibility between the requirements imposed on *any* law of nature by the theory of relativity (that is, the requirement of objectivity according to the principle of relativity, in its special or general form) and the formal mathematical structure of the equations of quantum mechanics, as a basic law of matter (see chapter ten for further discussion on this point).

If nothing but a single particle of matter existed in the

universe, then since this particle would not have any other particle to interact with, it would have to be described in an inertial frame of reference, represented precisely with special relativity. This matter could be described as being at rest (in its own frame of reference) or moving at a constant speed in a straight line relative to the reference frame of some observer (that exerts no force on it). But in practical terms, as soon as a second body should appear in the universe, *anywhere* relative to the first body, it would interact with it, causing it to move in non-uniform fashion, i.e. causing it to move *non-inertially* relative to the second body's position (and vice versa, in accordance with Newton's third law of motion). This situation is closer to the real world than the previous case. For even if there could be a single, localized body in the entire universe, with a *natural* (rather than supernatural) observer outside it, looking at it with instrumentation of any sort, the 'observer' would influence the 'observed', causing it to move non-uniformly (i.e. to accelerate) relative to the observer's frame of reference, no matter where the observer might be relative to the observed. This is because interactions between matter are generally of infinite range, *in principle,* in a relativistic theory, even though for practical purposes one may assume that there is a cut-off not far from the observed matter *as a particular approximation.*

We must then conclude that the model of a universe in terms of a superposition of individual things, each in an inertial frame of reference relative to the others, is not the real world. For in reality, each frame of reference relating to the localization of the matter components of the universe, or any small part of it, must be in non-uniform motion relative to each other such reference frame by virtue of the forces that are *mutually* exerted.

This argumentation then implies that the theory of special relativity may, *in principle,* only relate to a vacuous universe. The correct laws of any phenomena involving material components of the universe must then conform to the rules of the theory of general relativity: the requirement that the laws of nature are objective with respect to space-time transformations between reference frames that are in relative nonuniform motion. In this case, the invariant interval s^2 (equation (1)) does not summarize the transformations that leave the laws of nature unchanged. Rather, Einstein discovered that this must be altered to the constancy of a more complicated interval:

$$(3) \quad S^2 = g_{oo}(\mathbf{r},t)c^2t^2 + g_{11}(\mathbf{r},t)x_1^{\,2} + g_{22}(\mathbf{r},t)x_2^{\,2} + g_{33}(\mathbf{r},t)x_3^{\,2} + g_{01}(\mathbf{r},t)ct(x_1) + \ldots + g_{23}(\mathbf{r},t)x_2x_3$$

where the 'dots' above denote the other cross product terms. This is called a 'Riemannian space-time'. The *ten* coefficients, $g_{ab}(\mathbf{r},t)$, are functions of the space *(r)* and time *(t)* variables; their values depend upon where in space and time they are evaluated. The reason that there are exactly ten such coefficients is that they are symmetric, $g_{ab} = g_{ba}$. Thus, the ten independent coefficients are: g_{oo}, g_{11}, g_{22}, g_{33}, g_{01}, g_{02}, g_{03}, g_{12}, g_{13}, g_{23}. The collection of these coefficients is called the 'metric tensor'. Instead of a 'light cone', the geometric shape of the new S^2, of general relativity above, may be thought of as a 'squashed light cone'.

In spite of the fact that the logic of relativity theory implies that the real world, with matter, must necessarily be characterized by the invariance of S^2 of general relativity, rather than the invariance of s^2 of special relativity, it is still true that special relativity has been quite successful in correctly predicting a large number of empirical facts. The logic of relativity theory then implies that we must view special relativity as a good approximation, but still not an exact formulation, under those conditions where it has been successful. That is, it must be kept in mind that special relativity and general relativity apply to two different situations, one being the real world of matter and the other being the idealization of a vacuous world. Thus, to describe material forces in astronomy or in elementary particle physics with the rules of special relativity theory is fallacious if interpreted as an exact formulation. However, it is valid to do this if it is kept in mind that this is an *approximation* for a formulation in general relativity. Such an approximation then entails the following *asymptotic limits* of the components of the 'metric tensor':

$$g_{oo}(\mathbf{r},t) \rightarrow 1, \; g_{kk}(\mathbf{r},t) \rightarrow -1, \; (k = 1,2,3), \; g_{a \neq b}(\mathbf{r},t) \rightarrow 0$$

In this case, we would have the asymptotic approximation of

the (true) general relativity interval approaching the value of the special relativity interval, $S^2 \rightarrow s^2$. The reason that we must always keep in mind the fact that when we use special relativity it is a mathematical approximation for general relativity is that there are still some features of general relativity that should appear when special relativity is used, but could not be derived from special relativity itself. An example of this, in the framework of the theory of relativity in its general form, is the inertial manifestation of matter, when the latter is defined in accordance with the *Mach principle* (as we will discuss in chapter nine).

Why is it that in general relativity, the coefficients of the metric tensor, g_{ab}, depend on the coordinates of space and time, $\mathbf{r}$ and t? The reason is as follows: the space and time coordinates play the role in relativity theory of the words of a language that is invented for the express purpose of facilitating a formulation of the laws of nature. But a language involves a logic as well as words. The logic of the space-time language, analogous to the syntax of ordinary language, is in two parts: algebra and geometry. The *geometrical logic* of space-time refers to the relations between the points of space-time in the sense of relative congruence, parallelism, mapping, and so forth. The *algebraic logic* of space-time refers to the relations between the space-time points in the sense of rules of combination, countability, commutativity, etc. The essence of the algebraic logic of space-time is expressed most compactly in terms of what we call a 'symmetry group'—the *Poincaré group* for special relativity and the *Einstein group* for general relativity.

Now, since the relations between the space-time points and lines must reflect the nature of matter in the language of the laws of matter, and since matter is generally variable in space-time, it follows that the relations between space-time points in the space-time language must also be generally variable. But the relations between the space-time points of a Euclidean space-time, such as that characterized by the constant interval s^2 in equation (1), are the same everywhere. Thus, Euclidean geometry is not an adequate logic to represent the laws of matter. Einstein then went to the Riemannian metric of equation (3), where the relations between the space-time points are generally variable from point to point, as expressed in terms of the coefficients of the metric tensor, $g_{ab}(r,t)$. The reason that Einstein chose this particular non-Euclidean geometrical logic out of a few others that were at his disposal when he first studied the formulation of general relativity theory, was that the invariant metric S^2 also has the feature that it approaches the special relativity invariant interval, s^2 *in continuous fashion,* in accordance with the approximation that must be reached where special relativity has been empirically valid in physics. Recall again, however, that the *actual* limit of the description in special relativity is in principle unachievable since in an exact sense it only represents the state of the universe as vacuous. Nevertheless, it is important to keep in mind this boundary condition of *approach* to this limit, where we may use the equations of special relativity as an accurate mathematical approximation. This is an example of the imposition of the *principle of correspondence,* as we have discussed in previous chapters in regard to the way in which the formal expression of the quantum theory must approach that of classical physics in the appropriate limit (of large separations and action).

Summing up, then, all of the laws of nature that pertain to matter (including the formal expression of quantum mechanics) must be expressible within the framework of the theory of general relativity if they are indeed objective laws of nature. We have argued earlier that quantum mechanics may not be described within this framework in an exact sense. This seems then to lead us to the conclusion that quantum mechanics may not express a valid law of nature if the theory of relativity is also to be acceptable as a valid law of nature. Still, in accordance with *the principle of correspondence,* the formal expression of the quantum theory, in expressing the empirical facts about micromatter at low energy (that is, in the nonrelativistic approximation) in a mathematically consistent way, must be taken as an accurate mathematical approximation for an entirely different general theory of matter in the domain of elementary particles and greater—the theory of general relativity.

Space-Time Curvature and the Principle of Equivalence

One of the very important and interesting features of the

Riemannian space-time, characterized by the constancy of the interval S^2 (equation (3)) with respect to transformations to all space-time reference frames, is that any extremum path in this space-time, such as the shortest distance between any two points of the space-time, is a *curve,* rather than a straight line. Such an extremum path is called a 'geodesic'. The geodesics of a Euclidean space-time are the family of straight line paths. Such a space-time is called 'flat'. But the family of geodesics of the Riemannian space-time, *as observed from any frame of reference,* are a family of curves. Such a system is then called a 'curved space-time'. It should be noted that a 'curved space-time' is not something that can actually be visualized with the senses. It is a feature of the geometrical logic of space-time in general relativity. To see that it cannot be actually visualized in concrete terms, consider the analogy that can be visualized: the geodesics on the surface of a sphere, such as the Earth. These are the great circles, the circumferences (longitudes) that all have the same total length and the same center (the center of the Earth). For example, if one wishes to travel the shortest path from New York to London, it must be along a great circle of the Earth: this would go through Labrador, and far into the North Atlantic, in contrast to the path one would take drawing a straight line, on a flat map, from New York to London. But one may visualize this geodesic in this way: when looking down on the surface of the Earth, the geodesics would be convex curves. On the other hand, looking at the same geodesics from the inside of the Earth, they would be concave curves. In this case, we are still describing a Euclidean space. On the other hand, in a Riemannian space, *all* of the geodesics are convex (or all are concave), *from any frame of reference.* This is a feature of the Riemannian (curved) space that may not be visualized with our senses, though it can be fully described in a mathematical way.

It was Einstein's idea that if one could approximate a particular aspect of a closed system with an (approximately) disconnected test body, then the natural trajectory of this body must be a geodesic. If we were using a Euclidean geometrical logic, such geodesics would be the family of straight lines. This conclusion, for a Euclidean space, corresponds to the statement of Galileo's principle of inertia, asserting that the natural path of a free test body must be a straight line, at constant speed (as we have discussed in chapter one). On the other hand, if we should consider the 'free' test body to be in a Riemannian space-time, its (geodesic) trajectory would be a curve rather than a straight line, *as viewed from any frame of reference.* Not being aware of any geometrical system other than that of Euclid, if Galileo had seen the test body move along a curved trajectory, he would have claimed (in accordance with his principle of inertia) that it is not moving along a 'geodesic', but rather it is acted upon by some external influence (a 'force', as Newton called it a generation later).

Thus Einstein proposed that the action of one massive body on another, in causing it to accelerate in a particular way, such as the curved motion of a planet under the influence of the Sun, *is equivalent to* the 'free motion' of the body in a curved space-time. This is, in my view, the most general expression of Einstein's 'principle of equivalence'. It says that, in some way, the geometrical properties of space-time that have to do with its curvature are equivalent to the existence of forces that would act on a test body that moves along equivalent curves.

The next step that was necessary to substantiate this speculation was for Einstein to demonstrate the explicit relations between the geometrical properties of space-time, having to do with its curvature, and the forces that matter exerts on matter. This was demonstrated in the mathematical relations that Einstein discovered, called 'Einstein's field equations', that later on yielded the same predictions of the gravitational force as Newton's equations, in addition to new predictions that were not predicted by Newton's theory of universal gravitation.

Summing up, then, one of the spectacular successes of Einstein's theory of general relativity was his demonstration that the particular family of geodesics of a Riemannian space-time, *that correspond to the existence of the Sun,* according to Einstein's field equations, precisely predicted the planetary orbits of our solar system, in better agreement with the facts than did Newton's theory of universal gravitation. Einstein's theory then met the criteria to supersede the classical theory of gravitation because it not only reproduced all of the successful results of the classical theory but also predicted new effects having to do with gravitation that were not even predicted from a qualitative viewpoint by Newton's theory.

The major philosophic change that came from Einstein's theory replacing that of Newton was that the 'action-at-a-distance' theory, in which forces between distant interacting bodies act simultaneously, was replaced with the theory in which forces propagate at a finite speed (the speed of light c). The change also implied a replacement of the atomistic approach to matter with a continuum view, in terms of a theory of fields without singularities. From the mathematical viewpoint, Newton's equations are *linear,* in which the gravitational force between interacting bits of matter (e.g. planets, stars or galaxies) depends only on a single independent variable—the distance between them; whereas, Einstein's equations are *nonlinear* (implying that a linear sum of solutions of his equations is not generally another solution), and the terms that play the role of force in Einstein's theory, which relate to the curvature of space-time, generally depend on the four independent variables of this theory—the three spatial parameters and the time parameter.

Thus Einstein discovered that there are field equations that relate the geometrical properties of space-time, in terms of the 'metric tensor' field g_{ab} and its derivatives, to the matter fields variables (such as the terms that predict the energy and momentum of matter in any domain). Symbolically, Einstein's equations then have the form:

$$(4) \qquad G_{ab} \text{ (geometry)} = kT_{ab} \text{ (matter)}$$

where the left-hand side of this equation, G_{ab}, is called the 'Einstein tensor', and is a complicated (nonlinear) term depending on the metric tensor g_{ab} and its derivatives (rates of change in space and time); the right-hand side of this equation, T_{ab}, is the 'energy-momentum tensor' of the matter that is to be explained. In general T_{ab} is not 'energy-momentum' in the curved space-time, because there are no actual conservation laws in the curved space-time. But it is a function that approaches the actual energy-momentum of the matter considered as the space-time approaches flatness. This is a limit that does not exist in principle, for the reasons that we have discussed previously, though one may approach this limit as closely as desired: that is, to be an actual flat space could only correspond to a vacuum.[72]

Tests of Einstein's General Relativity

Not too long after Einstein discovered the explicit form of his field equations (4), and it had been found to reproduce all of the predictions of Newton's theory of universal gravitation, predictions were made of three additional effects that were not at all predicted by the classical theory. These new predictions had to do with the feature of his theory that the geodesic path of a test body relates to the density of a matter field and its intrinsic potential to interact with matter. It follows from this theory that the nonzero mass density in some vicinity of a test body must manifest itself as some nonzero curvature of space-time in that vicinity. This predicts that the test body's trajectory in that vicinity must be a particular curved path.

One of the three 'crucial tests' of Einstein's general relativity was the prediction that starlight, in passing the vicinity of the edge of the Sun, must bend in its path by a specified amount, as a function of the physical parameters of the Sun. (See figure 8.3). Knowing the position of the star from other observations at other places on the Earth and other times, the observation that the image of the star is where it isn't supposed to be then led to a measurement of the actual bending of the beam of starlight as it passed the rim of the Sun on its way to the Earth astronomer's telescope.

This effect was indeed observed (around 1920) during a total solar eclipse (so as to eliminate the blinding effect of the Sun's glare on the resolution of the star's position). It was observed then that the starlight did bend as it propagated past the Sun's rim, by an angle that was in very close agreement with the prediction of the theory. Since that time, new and better resolved measurements have been made, using the satellites of the U.S. space program, such as measuring the bending of a beam of microwaves (radar) as it propagated past the rim of the Earth, bouncing off of the moon, and then retracing its path by

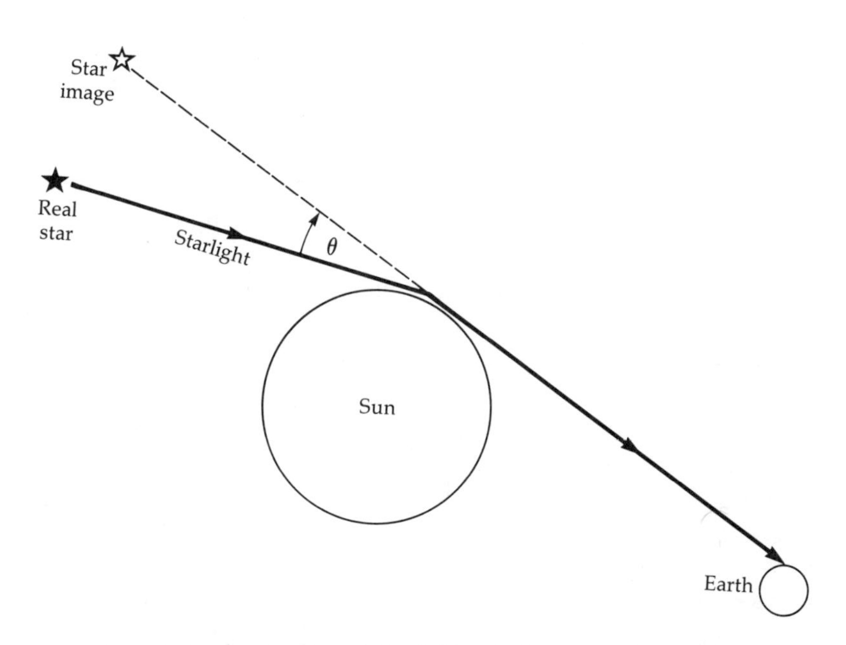

Figure 8.3. The Bending of Starlight as it Passes the Rim of the Sun

the rim of the Earth and back to the instrumentation in the satellite. These experiments verified the predictions of the theory of general relativity.

The second 'crucial test' of Einstein's theory had to do with the predicted effect of a massive body on the measured frequency of radiation in its vicinity—a type of measurement that amounts to a measure of time. Just as the spatial measure in the preceding example depended on the curvature of space-time, so does the temporal measure depend on it.

It then follows that the measure of a frequency of radiation must be influenced by the matter density of the domain from which it is emitted. The latter, in turn, is expressed in terms of the gravitational potential energy of the test body at the same place the emitter of the radiation is located. The prediction that follows from Einstein's theory is that the wavelength of the emitted radiation should increase with increasing gravitational potential. Thus, wavelengths of light in the visible range should shift toward the red end of the spectrum. The effect is then called the 'gravitational red shift'.

Such a shift in wavelength was indeed observed and

measured very accurately in the 1950's, when the technique of the Mössbauer effect became available to measure frequency shifts more accurately than had ever been done before. The experiment entailed gamma radiation on the roof of the physics building at Harvard University, measuring its wavelength and comparing it with the wavelength of similar gamma rays emitted in the basement of the same building. The experiments, carried out by R. Pound and his collaborators,[73] yielded remarkable agreement with Einstein's theory of general relativity.

Finally, the third of the 'crucial tests' of general relativity was a gravitational effect that was actually seen by astronomers in the middle of the nineteenth century, in regard to planetary motion. This was the observation that Mercury's motion about the Sun is not in fact periodic. And it was predicted by Einstein's equations that when applied to this case, the planetary motion should not be exactly periodic: the planet should take slightly longer each cycle to return to an equivalent position in its orbit, relative to the Sun's position.

The ideal type of planet to exhibit this type of effect would be the smallest one, since it would have the most elliptical orbit. It was observed in the middle of the nineteenth century by the French astronomer, Leverrier, that indeed the orbit of the lightest planet of our solar system, that of Mercury, is not exactly periodic. Focusing on the perihelion point of its orbit (this is the point of closest approach to the Sun, at one end of the elliptical path) Leverrier saw that each cycle it took a fixed time longer to reach the perihelieon of its orbit. (See figure 8.4.)

When this effect was seen, in the nineteenth century, it was at first noted that the effect could be due to the Newtonian forces exerted by the other planets of our solar system on Mercury, for it is the effect of the sun alone that predicts the periodic orbit. The Newtonian gravitational forces between Mercury and the other planets would cause an aperiodic contribution to Mercury's orbit. But when these perturbations on Mercury's motion were calculated from classical physics, they were found to account for only about 90% of the anomalous behavior. Yet, the errors in the measurements were much less than 10%. Thus, there must have been a part of the

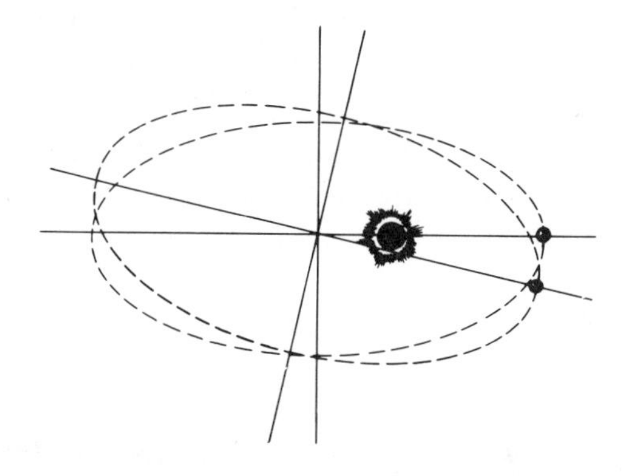

Figure 8.4. Deviation from Periodic Motion in Mercury's Orbit

aperiodic behavior of Mercury that was not explained by Newton's theory of universal gravitation. It wasn't until Einstein's theory of general relativity had appeared in the twentieth century that this 10% contribution was predicted by Einstein's field theory—in very good agreement with the experimental data.

Two of these three crucial tests of Einstein's general relativity were effects that had not been observed until they were predicted. The experiments, on the bending of light near the sun and the gravitational red shift, were spectacular successes for the theory. The agreement with the third test, the aperiodic contribution to Mercury's orbit, was not as spectacular to the physics and astronomy community as the other two because the theoretical result came after the experimental facts were known. But this test was certainly as important as the other two. The timing of experimental confirmations of a theory should have nothing to do with its significance for the scientific truth of that theory.

Successful as Einstein's theory of general relativity was, in superseding Newton's theory of universal gravitation, it could hardly yet claim such success beyond the relatively infinitesimal domain of the universe that is our own solar system. That is, it did not yet predict stellar motions within the galaxies, the motions and shapes of the galaxies, and the dynamics of the galaxies of the universe as a whole: the science of cosmology. Cosmology is both the oldest and the youngest branch of physics. It is the oldest from the point of view of speculation about the mapping of the stars of the heavens, such as the theoretical views of Pythagoras in ancient Greece. It is the newest from the point of view of the dynamics of stellar systems in terms of direct cause-effect relations as represented in modern-day astrophysics.[74]

Elementary Interaction

Einstein's theory of general relativity is not only a theory of gravitational phenomena, as many contemporary physicists contend. It is a general theory of matter, in all domains. This was Einstein's view because it is built into the logical structure of the theory, implying in principle that what we have so far is only the beginning of a unified field theory, in which all possible force manifestations and the inertial manifestation of matter are implicit in universal laws. Einstein spent the last 30 or 40 years of his life in seeking a unified field theory, not for purely esthetic reasons, but rather to complete the partial theory that had been discovered thus far, a completion that was *logically necessary* in order to actually claim the truth of this theory as a law of nature.

The theory of relativity is a law about laws, rather than a law about phenomena; it imposes the restriction of objectivity on the cause-effect relations that deal directly with the behavior of matter in fundamental terms. This theory implies the imposition of restraints on the laws of matter in all domains, from elementary particle physics to cosmology. The philosophy of this approach implies that we must start in principle with a closed system, rather than an independent thing. Thus it is the elementary interaction, rather than the elementary particle, that must be considered as the starting point.[75]

Elaborating on this point further, we see that according to the principle of relativity it should make no difference to the *objective* representation of a material system if one of the interacting components is called by the name 'observer' (the subject) and the rest called 'observed' (the object), *or vice versa.* Since the subject and the object of any elementary interaction are interchangeable without altering the fundamental description of the system, the approach of relativity theory to matter must be fully objective at the outset. This is to be contrasted with the Copenhagen approach in quantum mechanics, as a fundamental theory of matter, where there is indeed a degree of irreducible subjectivity in its basic description since there is not total symmetry between 'observer' and 'observed'. This fundamental difference is rooted, in fact, in the philosophic *realism* of the theory of relativity versus the positivism of the quantum theory.

The Allegory of a Dog-Flea World

The constancy of the 'observer-observed' relation with respect to reference frame, according to the theory of relativity, and its sharp contrast with the idea of the subjectivist, atomistic view of matter, according to the quantum theory, may be illustrated with the following allegory: consider the hypothetical world that is made up of a large, hairy dog, a small flea that lives on its back, and their environment of sky, Sun, ground, and so forth. From our reference frame, in looking in at this world, we may estimate that the dog has a lifespan of about 18 years and the flea's is about one day; thus we should expect them to have different senses of time scale, if they have such a sense at all. If we should then ask the dog what he knows about the world, he might start out by telling us what a handsome, strong and modest dog he is, with a good mind and a shiny, long tail and beautiful fur. Then he might go on to say that he lives on cool, still ground, and that he enjoys the bright days, with the warm sunshine. He would then go on to relate to us one of the troubles in his life: that he has a terrible recurrent itch on his back. Even though he scratches himself every few minutes, the itch comes back almost immediately. And then, every few hours, he scratches himself with his paw and shakes himself thoroughly; this makes the itch go away, only to return five or ten minutes later on another part of his back.

If we should ask the flea about his world, he might start out like the dog, telling us about his own self-attributes: that he is bright, with shiny wings and strong legs with sharp claws that can dig into the ground so as to make him secure when he thinks there may be trouble brewing. He might then go on to say that he lives in a forest of tall, soft trees, on undulating

Figure 8.5. Relativity and Objectivity. *Whose description of the world is correct, the dog's or the flea's?*

warm ground, rarely ever seeing the sky. But he would continue that he has the trouble that every few years of his life there is a terrible earthquake and a mountain descends to the ground near him, almost tearing him to pieces. The shaking ground usually throws him up into the air, only to land in a strange new place, though still on warm, undulating ground surrounded by tall soft trees.

Certainly, these are quite different stories about *the* real world. Who is telling the truth? The answer, of course, is that both the dog and the flea are telling the truth though each is telling an *incomplete* truth, since each is told from a subjective point of view.

If there is a single, real world, these two observers (and participants) in it must agree about its essential features. If the flea and the dog would take the philosophic position of the theory of relativity, their respective accounts of the world would follow the view of a *closed system.* The dog might start out just as he did before, though leaving out statements about the 'self' as objectively true facts. But he would go further by trying to *understand* his 'itch': the flea and its reactions to himself, as a dog. A close study might then reveal that there is a living thing on his back, why it causes him to itch, and so forth. Similarly, the flea would start his story as before, though also leaving out his ego-building statements. He would then go on to report on his study of his environment (the dog) finally understanding that there is this living creature that he lives on, with reactions to his digging claws that causes it to shake all over, and so forth.

The dog and the flea would each use different languages, different scales of space and time, and generally they would each indicate that they see the world differently in terms of their quite different respective modes of thinking. But if each of them *completes* its account of the world, it would be a world of *dog-flea,* a closed system, and each account would be in one-to-one correspondence. If each of them learned the language of the other and learned to communicate with him, he would realize that he is giving the same account of the real world as the other, though in different scales of space/time. This view would then be in accord with the theory of relativity, in which the *objective* account of the real world, the one that is independent of each participant's expression of it, corresponds to the objective laws of nature, and the dog's and flea's languages correspond to the different scales of space and time measures in the different reference frames.

From the point of view of understanding the behavior of matter in a fundamental way, the implication follows from the theory of relativity that the most elemental form of a law of nature must start at the outset with the notion that it is the whole that is elementary; that is to say, the whole is not a sum of separable parts. In the example above, the dog-flea closed system (where the hyphen between their names is irremovable) is the world. It is only when we take an asymptotic limit, corresponding to a weakening of the coupling between the (supposed) parts, that the system *appears* to be made out of 'parts'. But with the full logical standpoint of the theory of relativity, this would be an illusion just as it would be illusory to think of the ripples of a pond as independent component parts that the pond is made out of. For we know that the ripples of the pond are not more than *distinguishable manifestations of the whole, single pond.* In this same sense, any material component of the universe (a galaxy, a star, a planet, a human being, an electron) is only a distinguishable manifestation of the single universe, that is in principle without separable parts. This holistic view, of course, is in sharp contrast with the atomistic views of both the classical theory of matter and that of quantum mechanics of the present period in the history of physics.

A mathematical implication of this view of *oneness* is that in the precise formulation of the laws of nature, in all domains, from elementary particle physics to cosmology, there must be coupled equations for the material components of the system, in which, in principle, the coupling may never be off. This is because the material components we refer to are not really 'parts'; they are rather distinguishable *modes* of the single whole. Under particular, specialized circumstances, one may use the mathematical approximation for these coupled equations that makes it appear that there are separate parts that perturb each other. But it is important to always keep in mind the fact that the perturbation may never be truly off since it represents an approximation for a system whose reality is coupling. Mathematically, this corresponds to saying that there are no 'free field' solutions to start from. There is only the closed system that one may approximate (asymptotically) by assuming a weakly coupled system. The mathematical expressions of the coupled equations for such a system are 'nonlinear', and the fact that the coupling may never be off implies that there is no exact linear limit for these equations (though one may *approach* linearity). It is in this way that one recovers the linear structure of quantum mechanics from the nonlinear structure of general relativity. This view of the closed system in general relativity theory is highlighted in an interpretation of the inertial mass according to Mach, which is in conflict with the Newtonian interpretation of the particle views. This interpretation of inertial mass, called 'the Mach principle' was discussed in detail in chapter one.

Motion

In classical physics, 'motion' is characterized by the rate of change of spatial position of some quantity of matter with respect to a change in time. In contrast, in the theory of relativity one may not refer to such a rate in any objective sense because a purely spatial measure or a purely temporal measure in one reference frame would be expressed as mixtures of spatial and temporal measures in other reference frames that may be in 'motion' relative to the given reference frame. That is, what would be a length measure of one meter in one reference frame would have to be expressed as the combination of some spatial measure (say 80 **cm**) and some time measure (say 30 **sec**) in some other relatively moving frame. Thus, if the motion is characterized by *dr/dt* **cm/sec** in one reference frame, this parameter would be in terms of the rate of change of some space-time measure of one reference frame with respect to a different space-time measure of the other relatively moving reference frame, i.e. $dr \rightarrow dr' = \alpha dr + \beta dt$ and $dt \rightarrow dt' = \gamma dr + \delta dt$.

The term, 'motion', in relativity theory, then refers to a *general change* from the language of space-time measures of one reference frame to the space-time language of other space-time frames where one may wish to compare the forms of the laws of nature. Thus 'motion' refers to a 'translation' of the words of a language from one reference frame to another, rather than to some absolute attribute of some quantity of matter, as it is in classical physics and in earlier philosophical views of matter, such as in Aristotle and in Descartes.

One feature of motion that is implied by the assumption that the space and time parameters are continuously distributed is that motion must also be characterized by a set of continuous parameters. From the sense impressions that we have, as human beings, we are used to thinking that space and time are indeed continuous, and that motion is also continuous. But the philosophic position of abstract realism that underlies the interpretation of the theory of relativity teaches us that our sense impressions do not necessarily give us a true picture of the reality that may underlie these impressions. That is to say, there is no *a priori* reason to reject the concept of a discrete space, time and motion, if they can be defined in a rational way and lead to accurate expressions of the laws of nature. In this case, the laws would be 'difference equations', governed by algebra, rather than 'differential equations', governed by calculus. Nevertheless, to this point in time in the history of physics, there has been no reason to reject the concept of continuous space, time, and motion on the grounds of comparing the logical deductions following from laws in continuous space-time and the experimental facts.

Thus we see that the term, 'motion', in the theory of relativity, refers to the transformations of the continuous space-time languages of different reference frames in which one may wish to compare the expessions of laws of nature. This is analogous, in ordinary language construction, to the full set of translations of one language, in terms of words and their logical connections (syntax), into another language, such that the meanings of all possible sentences would be preserved.

Matter—Continuity Replaces Atomism

Assuming that motion is a continuous entity in the mathematical language of the laws of nature, it follows that the dependent variables that solve the laws of matter must also be continuous functions of the space-time parameters. Further, analysis requiring that among the laws of nature we must include conservation laws (of energy, linear momentum, angular momentum) leads to the additional restriction that the solutions of the laws of nature must not only be continuous in space-time, but that they should also have derivatives of all orders. Such functions are called 'analytic', or 'regular'. An analytic function has no 'singularities' in space-time, *anywhere*. This feature of the solutions of the laws of nature has profound implications with regard to the true nature of matter.

The first implication that follows is that indeed there are no localized atoms of matter—molecules, atoms or elementary particles—as singular things of matter. For if such entities did exist, the variables of matter would have to exhibit this atomistic feature of having a 'cut-off' in space, where the matter would cease to be, and the self-energies, confined to special spatial domains, would be discretely singular. What appear as atomistic features of matter, according to the field theory implied by relativity theory, are only a particular ('asymptotic') form of a continuous matter field, as the ripples of a pond are a manifestation of the continuous pond.

The Correspondence Principle and Special Relativity Transformations

An important feature that is built into the theory of relativity (in its special and general form) is the *principle of correspondence*. This is the assertion that the formal expression of any new theory in physics must (at least) approach the formal expression of an older theory that it claims to supersede, in the limiting form of the new theory that would apply to the domain where the older theory had been empirically successful. Thus, as appropriate parameters become small, the predictions of the theory of relativity should agree with those of classical physics.

Galileo discovered that the laws of motion, that he expressed 'kinematically', using the language of geometrical concepts, must correspond in regard to their expressions in relatively moving inertial reference frames, that is, frames that move with respect to each other at constant relative speed in a straight line. Though these were not his exact words, this is the essence of Galileo's principle of relativity. According to Galileo, if one reference frame is in motion at a constant speed, v, in a straight line, relative to some other frame of reference, then the laws of nature, such as the law of a 'freely falling body' (that it accelerates at a constant rate, independent of its mass) should be the same if expressed in any other inertial frame of reference. If the relative motion between reference frames is defined to be in the x-direction, then the transformations of the coordinates between them, to preserve the laws of nature, are in Galileo's theory,

(5a) $$x' = x + vt, \; y' = y, \; z' = z,$$

Since there then seemed to be no question about the absoluteness of the time measure, it was assumed that

(5b) $$t' = t$$

Galileo's principle of relativity then asserts that the laws of nature must be the same in all reference frames that relate to each other in terms of the space and time transformations shown in equations (5a) and (5b).

Three hundred years after Galileo, Einstein claimed that the laws of electromagnetism (in particular) are not preserved under Galileo's transformations (5), but are unchanged under the *Lorentz transformations* (for the case of constant relative motion in the x-direction) as follows:

(6a) $x' = (x + vt)/[1 - (v/c)^2]^{1/2}, y' = y, z' = z$

(6b) $t' = [t + (v/c^2)x]/[1 - (v/c)^2]^{1/2}$

Here as before, v is the relative speed in a straight line, in the x-direction parallel to the x' direction, and c is the speed of light in a vacuum. Einstein concluded from this that these transformations must apply to all of the laws of nature when their forms are compared in relatively moving inertial frames of reference.

Note that it is then the set of transformations (6a) and (6b) and not (5a) and (5b) that also preserves the constancy of the invariant length s, as defined in equation (1). Thus the constancy of s is a compact way of summarizing the set of all space-time transformations that preserve the forms of the laws of nature. When the relative motion that distinguishes the reference frames is inertial, then this theory is referred to as 'special relativity'. (When the reference frames are not inertial, for example when they are accelerating with respect to each other, then the theory refers to 'general relativity').

The *Lorentz transformations* (6a) and (6b) then show how the spatial and temporal measures 'mix', when the relative motion is in the x-direction colinear with the x'-direction. Thus, if some law of nature in the K-frame of reference (that uses the coordinates x,y,z,t) has the general form

$$O(x,t)F(x,t) = G(x,t)$$

then in another inertial frame, K', this law should have the corresponding form, according to the principle of relativity:

$$O'(x',t')F'(x',t') = G'(x',t')$$

where the transformations of space and time measures (6) predict how O, F, and G transform into O', F' and G'.

If the Galilean transformations (5) had been applied to the law of electromagnetism (Maxwell's field equations) it would be seen that these laws do not preserve their objectivity: they would have a different form in different inertial frames of reference. Since the space and time transformations must apply to all of the laws of nature, according to the principle of relativity (of either Galileo or Einstein) it follows that Galileo's transformations (5) must be false. But since they were adequate to describe Galileo's observations, such as objects sliding down inclined planes or the motion of a planet in its orbit about the Sun, the *principle of correspondence* requires that the space and time transformations (5a) and (5b) must be a good mathematical approximation for the Lorentz transformations (6a) and (6b), under the physical conditions that prevailed in Galileo's experiments. That is, if $v/c \simeq 0$, then equations (5a) and (5b) $\equiv$ equations (6a) and (6b).

This is clearly the case, since the speeds, v, that Galileo observed, were always very small compared with the speed of light, c. The largest speed he observed was that of the planet Mercury when it is closest to the Sun, relative to the frame of reference of the Sun. This speed is the order of 10^7 **cm/sec** (which is about $10^{-3}c$). In this case, the pertinent parameter in the transformations (6a) and (6b) is $(v/c)^2 \cong 10^{-6}$. Thus, in using his transformations (5a) and (5b) Galileo was making an error of only one part in a million, an error that was certainly beyond the resolving power of his instrumentation. Thus, he would have had no reason to doubt the authenticity of his space-time transformations (5a) and (5b).

The statement of Galileo's principle of relativity then remains as a principle of nature. It is only that the forms of the transformations of the space and time measures had to change, from those shown in equation (5) to equation (6) in progressing from Galileo's to Einstein's principle of (special) relativity.

In the next chapter we will discuss some well known paradoxes that have appeared in the history of Einstein's relativity theory, when the erroneous interpretation was used which says that instead of language translations, the transformations (6) are physical changes of material extension or temporal duration. We will then go on to the meaning of matter/antimatter and radiation in a purely relativistic field theory (in Einstein's sense). Finally, in chapter ten, we will summarize the basic dichotomies that are encountered in any attempt to fully fuse the quantum and relativity theories, and some discussion will be given to possible resolutions of this most important conflict of ideas in physics.

Nine

Further Implications of Relativity Theory

The Fitzgerald-Lorentz Contraction and the Clock Paradox

According to the interpretation of the Lorentz transformations (8.6) (equations (6), chapter 8), held by most contemporary physicists, the *physical extension* of an object in motion is relative to its state of motion with respect to any observer. That is, the claim is made that the physical length of a travelling meter stick would be less than a meter, from the viewpoint of a stationary observer. Similarly it is claimed that because of the scale change in time measure, according to the Lorentz transformation formula (8.6), durations of moving objects are also less than would be the duration of an identical physical object in the observer's frame of reference.

But this type of conclusion is based on an interpretation of the Lorentz transformations that is contrary to the view of the role of space and time in the theory of relativity: that they are not more than the words of a language whose only purpose is to express a physical law, *but they are not physical entities in themselves.* Indeed, it is the physical laws, and not the language that is used to express them, which govern the cause-effect relations that predict physical effects, such as the shortening or elongation of a material stick, or the extension of the duration of some physical process, such as the aging of a human body or the rotation of the hands of a clock.

The space and time transformations in Einstein's theory of relativity are only meant to provide an observer in a particular frame of reference with the proper rules for translating his expression of the laws of nature into their expression in any other reference frame. This is entirely analogous to the translation of English into French in expressing some sentence with a particular meaning. The translation is not meant to alter the meaning of the sentence. If there should be a physical change in a body that is in motion relative to a 'fixed' observer, *because of this relative motion,* then a logical paradox clearly ensues. This is because, motion, per se, is strictly a subjective term. One may say that some reference frame, A, is in motion relative to another reference frame, B, that is considered stationary; however, it may equally be said that it is A that is stationary and B is in motion relative to it. It does not matter if the motion is inertial (constant relative speed in a straight line) or not (non-uniform relative velocity, etc.) 'Motion' is a term that depends on the frame of the observer who claims to be stationary relative to it. Since 'motion' is a subjective part of the description of matter, it cannot imply any absolute physical consequence.

Let us pursue this somewhat further. If we consider the two ends of a meter stick at the same time, then according to the transformation formulas of special relativity theory (8.6), if the stick should move along the direction of its length, say in the x-direction, at the speed v relative to some stationary observer, then the stationary observer would use the following *scale change* to represent the measure of the stick's length in the moving frame relative to his:

$$\Delta x = \Delta x'[1 - (v/c)^2]^{1/2} \qquad (1)$$

where $\Delta x'$ is the measure of the extension of the stick in the stationary frame of the stick. Thus, the observer must use a contracted scale of spatial measure in the moving frame, where the contraction factor depends on the magnitude of the relative speed v compared with the speed of light, c. This is the 'Fitzgerald-Lorentz contraction'. But does this contracted scale of spatial measure really imply that there is a physical change in the extension of the stick? I will show that it cannot imply this if we are to maintain logical consistency.

The logical problem enters when we recognize that motion is a subjective part of the description while physical changes cannot be subjective. If we say that a meter stick L_1 moves away from meter stick L_2, and as a consequence L_1 is physically shortened compared with the length of L_2, it may equally be said that the meter stick L_2 is the one that is moving, relative to L_1 (and in the opposite direction to the motion described from L_2's frame). Thus, according to the principle of relativity itself, which requires that the laws look the same from all reference frames, we must say that the claims made from both reference frames, that of L_1 and that of L_2, are correct. Thus we conclude that if there is physical change by virtue of motion, L_2 both shortens and does not shorten, relative to L_1. But this is indeed a logical paradox. If the theory of relativity is to be acceptable as a scientific theory the paradox must be removed. To do this one removes the faulty interpretation of the Lorentz transformation (1) as referring to a bona fide physical change in the material extension of the stick. In fact, the theory of relativity does not imply this interpretation at all. As discussed above, the transformations of space and time measure do *not* denote physical changes within the logical structure of this theory. They only denote scale changes in the measures of space and time intervals. *The main fallacy then is to confuse the measure of a thing with the thing itself.* For if a measure of a thing is not the thing itself, then the need for a contracted scale of measure does not imply a physical contraction.

This is the very same paradox that is encountered if one should insist that the time transformations of special relativity (8.6), when there is no spatial separation in one frame and the other,

$$\Delta t = \Delta t'[1 - (v/c)^2]^{1/2} \tag{2}$$

refer to a real physical change of duration of a material object.

With this interpretation, a twin brother who may travel into outer space for about two days, according to his own watch and his metabolism, would leave his brother on Earth to age physically in much more of his time and metabolism; he may age, say, 80 years in this period. With this interpretation, then, when the twin brother who took the trip returns home to his brother, meeting in the same inertial frame that he left from, say the family room of their home, he would be smooth skinned, clear thinking, standing erect with a full head of unchanged hair and all of his teeth. But his brother, who stayed at home, would be wrinkled, senile, bent over, with little gray hair left and no teeth; he may be about 102 years old, *physically,* while the travelling brother is only 22 years old, *physically*. They would be physically different, one very old and the other still quite young, even though both brothers were born at the same time from the same mother.

But just to say that such a thing happened because of the trip that one of the brothers took, would not be a logical paradox, if it were not for the fact that the physical (absolute!) change between these twins took place because of relative motion. That is, if the traveller had been the one to claim that he was stationary, seeing his twin brother in motion (in the opposite direction), then it would be the brother who remained on Earth who didn't age while the traveller would have aged about 80 years. According to the principle of relativity, both brothers must be telling the truth; thus after the round trip journey is over we would have to conclude that one of the brothers is both older and younger than the other brother. This is a bona fide logical paradox. It is commonly referred to as 'the twin paradox' (or, equivalently, the 'clock paradox', when we talk about physically identical clocks, initially synchronized and going out of synchronization because of their relative motion). This logical paradox implies, as in the analysis of the Lorentz-Fitzgerald contraction, that we must reject the interpretation of the space and time transformations of this theory as physical changes.[76]

Of course, any *logical* paradox is inadmissible in a bona fide scientific theory. Note that the dictionary gives two meanings for the word 'paradox'. One is that it refers to an unusual, unexpected result. This is not a logical paradox. The paradox we have been referring to here and previously in the text, is a *logical* paradox, such as: the man is both 22 years old and he is 102 years old, simultaneously. As in our previous conclusion about spatial measures in relativity theory, there is no paradox when we interpret the transformations of space and time measures as intended in the theory: strictly as 'language translations' when transforming the expression of a physical law in one reference frame into the language of another reference frame. That is, the Lorentz transformations are *not* 'cause-effect' relations, therefore they do not imply physical changes. This is similar to saying that if one should contract the time scale on the face of a clock, calling each of the numbers, 1/2, 1, . . . , 6, in going around once, this could not mean that anything strange happened to the workings of the mechanism that makes the hands go around on the face of the clock.

If one should nevertheless insist that the Lorentz transformations do imply physical changes, it must mean that the theory of relativity is *false,* as a scientific description of real matter. Indeed, many of the critics of Einstein's special relativity, such as Herbert Dingle, have used this reason to claim that this is a false theory. Dingle asks the question: Which of the relatively moving clocks is the one that is slow compared to the other?[77] Not arriving at a logically sound answer, he concluded that the theory of relativity is false. But his conclusion is also false because he tacitly assumes (with the rest of the physics community) the interpretation of the Lorentz transformations in terms of physical change. If the rest of the physics community is correct in this interpretation, then Dingle would be correct that the theory is false (according to what almost everyone says its transformations mean). But the physics community and Dingle are not correct in this, as it leads to a logical paradox and the logical basis of the theory itself rejects this interpretation. In point of fact, the interpretation of the Lorentz transformations in terms of physical effects was the one that Lorentz himself took. To him, it represented the physical effect of the ether on all measuring rods and clocks. But this view was rejected with Einstein's theory of relativity.

As I have emphasized earlier, the space and time measures in relativity theory are not 'things in themselves'. They are only language elements, whose sole purpose is to provide a means of *expressing* the laws of nature, that is, laws of matter. The Lorentz transformations, in turn, are the 'scale changes' of spatial and temporal measure, that ensure the objectivity of the natural laws. To explicate this point further, the spatial and temporal scale changes in this theory may be compared with the expressions of two men looking at each other as they move further and further apart. Suppose that each of these men is 6 feet tall, measured when standing next to one another. Each of them then sees the other getting shorter as they part because the angle subtended by the light rays from each to the other's eyes becomes progressively smaller. When they are sufficiently far apart, each of these men would appear as a point to the other, if there is an appearance at all. But it is still an *absolute* fact of nature that each of these men is 6 feet tall. Of course, this is not an exact comparison with the interpretation of the Lorentz transformations because the latter are not observable in the first place. A twin brother who takes a trip in a rocket at high speed does not 'see' his brother on Earth growing old while he remains a youth, even as an illusion, or vice versa. Nor would we see a meter stick shrink (even as an illusion) if it should move at a great speed past us. At least, the theory of relativity does not explicitly predict such an illusory observation.

What we should see must be based on 'physical cause-effect relations' which are prescribed by the laws of nature. *It is only these cause-effect relations that predict physical changes.* The scale changes in spatial and temporal measures in special relativity theory are as disconnected from physical changes as are the rules of translating Chinese into English, or vice versa, disconnected from the actual meanings of the sentences that are expressed in these languages.

With the space-time language of relativity theory it is possible to rigorously prove, in mathematical fashion, that the theory of relativity in itself *does not* predict the physical consequence of asymmetric aging, as we discussed above in connection with the 'twin paradox'. In a part of my research program I have shown this result in the following way: consider an arbitrary geodesic path in space-time, between the specific points s_1 and s_2.[78] (Recall that such a path is the one of minimal (or maximal) separation between s_1 and s_2). This path is determined generally in general relativity theory (discussed in the preceding chapter). The equation of motion of a test body is the equation of this geodesic. The terms in the equation that play the role of a 'force' (the cause of the path to be what it is) are geometrical terms (called 'affine connections'). The latter, in turn, entail the total influence of all of the matter surrounding the 'test body' on the path of this body. The source of these geometrical fields is called an 'energy-momentum tensor field', T_{ab} (1). Once the Einstein equations are solved, T_{ab} (1) yields its solutions, g_{ab} (1). The latter in turn are used to construct the geometrical fields that prescribe the path of body 1. But if there were a different energy-momentum tensor source, T_{ab} (2), a different geodesic field would be predicted from Einstein's equations, g_{ab} (2), that in turn would yield a different geodesic path.

Suppose that we then consider two such *different geodesic paths,* though starting and ending at the same space-time points, s_1 and s_2. They may be called P_1 and P_2. Let us now calibrate the face of a clock that is to follow path P_1 from s_1 to s_2, choosing the measure to correspond with the physical duration of the particular mechanism of the clock, such as the unwinding of its spring, if it is an old fashioned type of clock. Suppose now that we have a *physically identical* clock, and calibrate its reading exactly in the same way as the calibration of the first clock. Now let the first clock be a test body, moving along path P_1 from s_1 to s_2, due to the imposition of the energy-momentum tensor field T_{ab} (1). And then let the second, physically identical clock move from the same starting point s_1 to the same stopping point s_2, though along the different geodesic, P_2, due to the imposition of the different energy-momentum tensor, T_{ab} (2). The total aging of the first clock must then *correspond to* its total path length, P_1, in going from the space-time point s_1 to the space-time point s_2. Similarly, the total aging of the second clock during its evolution from s_1 to s_2 must *correspond to* its path length P_2 between the same two end points.

What I have shown in my research program is that when one uses the *most general* expression for the geodesic in general relativity, a form that is consistent with the full requirements of the geometrical *and* the algebraic logic of this theory, it follows *rigorously* that

$$P_1 = P_2$$

It then follows that the total aging of the clock that moves along P_1 is equal to the total aging of the clock that moves along P_2. Thus it was concluded that the theory of general relativity (and also special relativity, as this is a special limit of general relativity) does not predict any effect of relative motion upon aging, no matter what sort of motion it may be—inertial, accelerated, and so forth.

In the case of the twin problem, the initial space-time point and the final space-time point, s_1 and s_1, are, respectively, the family room of their home at some earlier time and the family room at some later time, when the traveller left on his trip and returned. The geodesic of the stay-at-home brother was determined by his energy-momentum tensor T_{ab} (1) corresponding to all of his circumstances on Earth. The geodesic for the travelling twin was similar to that of his brother on Earth, except that his energy-momentum tensor T_{ab} (2) included the circumstances of his being on the rocket and making the round trip journey, thus predicting a different geodesic path between the same two end points. The equality of their path lengths then implies that their physical aging was precisely the same during the time of the trip, as expected from the logical analysis of the theory.

Note that the assumption was made in this analysis that the two physical mechanisms were identical in all ways. If there were unforeseen forces acting on the cells of one of the twins that were not acting on the cells of the other, such as to age one of them differently from the other, we might have expected a difference in their physical ages after the flight was completed. This conclusion would have been based on cause-effect relations, not on the translation of languages.

Summing up, then, the space-time transformations of the theory of relativity, in terms of its logical meaning, do not predict any 'twin paradox' (or 'clock paradox'). Thus the theory of relativity remains logically consistent, and therefore scientifically sound.

Relative Simultaneity

One other concept of the theory of relativity that I believe led to a great deal of confusion, because of a lack of examination of the logical content of the theory, is that of 'relative simultaneity'. Mathematically, according to the Lorentz transformations of the time measure, its value in one reference frame must be expressed in any other inertial frame as a linear combination of a time measure and a space measure (the fusing of space and time into space-time). Again in the special case where two inertial frames move relative to each other in the x-direction colinear with the x' direction, with the relative speed v, then if Δt, Δx are time and space intervals in one frame and $\Delta t'$, $\Delta x'$ are the time and space intervals in the other (relatively moving) frame, then,

$$(3) \qquad \Delta t' = [\Delta t + (v/c^2)\Delta x]/[1 - (v/c)^2]^{1/2}$$

According to this equation, if two events should take place simultaneously, in an observer's frame of reference, so that $\Delta t = 0$, they would not be simultaneous in the relatively moving inertial frame, i.e. $\Delta t' \neq 0$. That is, in this case,

$$\Delta t' = (v/c^2)\Delta x/[1 - (v/c)^2]^{1/2} \neq 0$$

In accordance with the interpretation of the space-time measures as 'language elements' rather than physical consequences, this nonsimultaneity should only be interpreted in an abstract sense. It should be understood to denote a change of language elements for particular time intervals, to be substituted into the *expression of* a law of matter that, in turn, leads to predictions of physical events, as described by an observer who studies phenomena in a frame of reference that is in motion with respect to his reference frame.

To demonstrate the difference between a 'language change' according to the space-time transformations of relativity theory, and a real physical change, consider the following situation in special relativity: A dog is crossing the street at an intersection. As he steps down toward a manhole cover in the middle of the intersection, a worker standing nearby pulls a lever that causes the manhole cover to open. An observer standing at the corner would see the dog disappear below the street. Other observers on the corner would notice that the dog does not reach the other side. They would all conclude that the dog stepping down to the manhole cover and the worker pulling the lever to open the manhole cover must have been *simultaneous*. This was an absolute set of two events at the same time. It had an absolute consequence, that is, independent of who was observing it, the dog did wind up below the street.

Suppose now that a helicopter pilot, flying over this road, near the speed of light, also observes the dog crossing the street and the worker with his hand on the lever. From the signals he receives from this observation he might deduce that the two events (the dog stepping toward the manhole cover and the workman pulling the lever) are *not simultaneous*. He might then anticipate that the dog would keep walking to the other side of the street. On the other hand, the pilot would *see* that the dog does not reach the other side of the street. If he is scientifically inclined, the pilot might then try to analyze the situation by trying to figure out why he thought the dog would reach the other side of the road, whereas he seemed to disappear along the way, perhaps falling into the open manhole. He might then realize that he came to the wrong conclusion because his observations were made from a moving body, relative to the reference frame of the dog and the Earth. The latter is, after all, the significant frame of reference to describe the events, because if the dog does fall below the street, it is due to its interaction with the Earth, irrespective of who happens to be looking at it, from one frame of reference or another. To predict the correct outcome of events, the pilot might then 'transform away' his own motion parameters relative to the frame of the dog and the manhole cover, thereby projecting himself into the reference frame where the physical interaction takes place: the dog-Earth frame. He would then correct his original assertion, now claiming that the dog stepping toward the manhole cover and the workman pulling the lever to open it are physically simultaneous events, and that the dog would fall below the street. He would thus account for his not seeing the dog reach the other side of the road.

In coming to the correct conclusion, the pilot saw that to remove his own motion from the description, he had to use a particular transformation, which was in fact the Lorentz transformation (3). At first he might have thought that the simultaneity of events was relative to the reference frame. And indeed this is so as far as the language description for simultaneous events is concerned. But there is only one *physically* simultaneous set of events, that refers to the 'proper frame'—that frame in which the participants of the interaction itself are described. There is, after all, one objective conclusion for this situation, which has to do with physical simultaneity, while there is a relative simultaneity in *all* language expressions of the physical laws from the relatively moving frames of reference.

Matter and $E = mc^2$

One of the most remarkable predictions that came from Einstein's theory of relativity, in its special form, was the relation between the total energy of a quantity of matter and its intrinsic inertial mass, expressed by the formula, $E = mc^2$. The result implied that if a particle of matter with mass m is not in motion, and thus has no kinetic energy, and if it is not subject to any external potential, and thus has no potential energy, it must still have a residual energy equal to mc^2, where c is the speed of light (in a vacuum, of order 3×10^8 meters/sec.[79]). To compare this quantity of energy with ordinary energies that we experience in everyday life, we note that they are the order of the typical values of kinetic energy, $mv^2/2$, where the speeds v could be the order of 10^2 or 10^4 or even (in the case of the speed of Mercury relative to the Sun's position) 10^5 meters/sec. Still, the ratio of the matter's intrinsic energy to the highest of these is $(c/v)^2 \sim 10^6$. Thus, the intrinsic energy of matter is the order of millions of times greater than ordinary energies of matter.

This latent energy in matter must then be a tremendous potential as an energy source, not suspected to exist before the early decades of the twentieth century. It was not observed, empirically, before the onset of the theory of special relativity because in all ordinary experiments that measure energy, one actually measures the differences between two quantities (this is the 'transferred energy' that is involved). These differences, in turn, are between two quantities that each have the rest energy in them, thus they cancel out the difference. However, in the experiments in nuclear physics from the early decades of the twentieth century, in which the radioactive decay of a nucleus into smaller parts was studied, it was seen that the sum of the masses of the smaller resulting nuclei was not as great as the mass of the original nucleus. It was then discovered that the difference was exactly in the *binding energy* of the whole nucleus, in the form ΔMc^2, where the decay was

$$M \rightarrow M_1 + M_2 + \ldots$$

and

$$\Delta M = M - (M_1 + M_2 + \ldots) \neq 0$$

The binding energy ΔMc^2 is then converted into the kinetic energy of the disintegrated parts and radiation. Thus, Einstein's prediction of the relation of mass to intrinsic energy was well-confirmed.

Let us now discuss the meaning of this energy-mass relation. One idea that it does *not* mean is that 'energy *is* mass'. That is, the relation $E = mc^2$ does not mean that mass is identical with energy.[80] For the concept of inertial mass is *defined* as the resistance of matter to a change of state of its motion and the concept of energy is *defined* as the capability of the matter to do work. These are entirely different concepts. Thus one may not say that energy is mass or that mass is energy. To illustrate this, we note that there are domains (according to the theory of general relativity) where energy is not defined, where mass is defined. That is, in the global domain in general relativity theory, there are no conservation laws—laws that say that certain quantities relating to matter (energy, momentum, angular momentum, and so forth) are constant in time. But the inertial mass of matter is defined in a global sense in general relativity. This feature alone implies that one may not say that mass and energy are identical.

What should be said of the energy-mass relation is the following: deriving the inertial mass of matter from global considerations in general relativity, one may then take the *local limit* of its form (assuming sufficiently small distances and quantities of energy transfer that special relativity considerations may be used with accuracy, as an approximation), thereby arriving at an averaged 'mass field' in the local domain, called m. One then inserts m into the relation, $E = mc^2$, to determine the ability of this matter to do work, *in the local description*, that is to evaluate its 'rest energy'.

Thus we see that the relation $E = mc^2$ is, logically, an 'if-then' relation: *if E then m;* but it is not generally true that *if m then E,* since m is defined under circumstances where E is not defined.

The quantity in special relativity called mc^2 is referred to as the 'rest energy' because when the matter is in motion, with speed v relative to some observer's reference frame, the expression for its energy is:

$$(4) \qquad E = mc^2/[1 - (v/c)^2]^{1/2}$$

The state of rest then corresponds to setting $v = 0$, thus recovering the rest energy, mc^2. If we wish to take account of the finite value of the speed of the matter relative to an observer, but assume that v is very much smaller than the speed of light c, one may expand the denominator in the energy expression (4) in a binomial series, giving:

$$(5) \quad E = mc^2[1 + v^2/2c^2 + 3v^4/8c^4 + \ldots] = mc^2 + mv^2/2 + (v/c)^2 3mv^2/8 + \ldots$$

The first two terms in this series are simply the sum of the rest energy and the kinetic energy (as specified in classical physics). If v/c is a small number, it would then be safe to ignore the third term and all higher terms of the series, since they depend on successively higher powers of the very small number v/c. This is the 'classical approximation' for the energy of the matter in motion (and not subject to any potential energy), which still must include the rest energy term, mc^2, that is normally *much greater* than all of the other terms of the series. From this expression we see that when we are subtracting two energies that correspond to two different speeds of the matter, the rest energy term would cancel in the result, as indicated above.

Momentum in Special Relativity

The expression for the momentum of a particle of matter with mass m and velocity $\mathbf{v}$, according to Einstein's theory of special relativity, is as follows:

$$(6) \qquad \boldsymbol{p} = m\boldsymbol{v}/[1 - (v/c)^2]^{1/2}$$

The classical expression for the momentum of a particle is $\boldsymbol{p} = m\boldsymbol{v}$. Clearly, if the relative speed of the mass, v, is much less than the speed of light, c, the relativistic expression reduces to the classical expression. However, if one should wish to express the relativistic momentum in a form that *looks like* the classical expression, we may write equation (6) as

$$\boldsymbol{p} = m'\boldsymbol{v}$$

where

$$m' = m/[1 - (v/c)^2]^{1/2}$$

With this re-expression, it is important to note that the parameter m' is not the inertial mass *per se*. Rather, it is a parameter that depends on the inertial mass of the matter, m (it is m, not m', that represents the resistance of the matter to an external force) and it depends on the speed of the matter, v, relative to the observer. Thus, one should not say that as the speed of the body increases relative to the observer, its inertial mass increases in the way that m' depends on v. Rather, it should be said that the momentum of the particle of matter, p, increases with v as the relativistic formula (6) indicates.

Note that the 'momentum' of the body is a relative concept, since it is defined in terms of the speed of a body relative to a fixed observer. If the observer should 'ride with' the body, it would have no momentum as far as this observer is concerned. An important thing about momentum is that its change in time is another way of expressing the force that acts on it. For example, if a proton should move at a speed close to the speed of light, c, it would have a very large momentum relative to some target nucleus, according to equation (6) (much larger than the classical momentum, mv). If it should then be stopped in a very short time, say by being embedded in nuclear matter, then its momentum change in time, and thus the force it imparts to the embedding nucleus, could be extremely large, perhaps large enough to break apart this nuclear target. This was the purpose of the original particle accelerators in nuclear physics experimentation: they were used to try to disintegrate complex nuclei so that the fragments could be studied (unbound neutrons, protons and mesons) in an attempt to learn something about the forces that originally bound them.

Radiation from a Moving Source—The Doppler Effect

The theory of relativity had some important successes in its predictions about the observations of monochromatic radiation (radiation with a fixed frequency) from sources that are in motion relative to an observer (or vice versa). Formally, when the charged matter source terms in the theoretical description of the electromagnetic equations are set equal to zero, the field solutions of Maxwell's equations yield the result that electromagnetic radiation propagates as a transverse wave. That is, the electric and the magnetic fields of force are vibrating in a plane that is perpendicular to the direction of propagation of the wave. The spectrum of the possible frequency solutions then corresponds to the radiation solutions of Maxwell's equations (discussed briefly in chapter two). The frequencies of these waves relate to their wavelengths in terms of the relation

$$\nu = c/\lambda$$

where the speed of propagation of the wave is the speed of light, c.

It follows from the objectivity of Maxwell's equations with respect to their expression in different inertial frames that the electromagnetic wave depends on the following function of space and time coordinates:

$$\exp(i\phi) \doteq \exp[i(k_o x^o - \boldsymbol{k}\cdot\boldsymbol{r})]$$

where the 'phase' term, ϕ, is a scalar, i.e. it is the same in all inertial frames of reference. The parameters that appear in this phase are:

$$k_o = 2\pi\nu/c,\ x^o = ct \quad \text{and} \quad /\boldsymbol{k}/ = 2\pi/\lambda$$

Because the phase ϕ is a scalar in special relativity theory, the components (k_o, k_x, k_y, k_z) must transform from one inertial frame to another in the same way that the components of the four-vector, (x^o, x, y, z) transform. That is, they must transform in accordance with the Lorentz transformations (8.6) of the preceding chapter. The transformed components of this 'four-vector', called k_μ, from one inertial frame to another, are then as follows:

$$k_x' = [k_x - (v/c)k_o]/[l - (v/c)^2]^{1/2},$$
$$k_y' = k_y,\ k_z' = k_z,$$
$$k_o' = [k_o - (v/c)k_x]/[1 - (v/c)^2]^{1/2}$$

where $k_o = 2\pi\nu/c$ and $k_x = /\boldsymbol{k}/cos\theta$, where it is assumed that one inertial frame is moving in the x-direction, parallel to the x' direction of the other frame, at constant speed v, ν is the frequency of the wave (in **Hertz**), and the magnitude of its wave vector is $/\boldsymbol{k}/ = 2\pi/\lambda$, where λ is its wavelength. The angle θ is between the direction of propagation of the wave and the direction from which the wave is observed. It then follows from the last of the equations above that the comparison of the measured frequencies in the different inertial frames is as follows:

(7) $\nu' = \nu[1 - (v/c)cos\theta]/[1 - (v/c)^2]^{1/2}$ **Hertz**

The result that the frequency measured in the moving frame of reference is not the same as the frequency of the radiation in the frame of reference of its source is called 'the Doppler effect'.

In the case where the observation would be along the same direction as that of the propagating wave (which is the direction of the wave vector $\boldsymbol{k}$), we would have to insert $cos\theta = cos0° = 1$ into equation (7), yielding the 'longitudinal Doppler effect',

(8) $\nu' = \nu[(1 - v/c)/(1 + v/c)]^{1/2}$ Hertz

This result then implies that if the source of radiation is moving away from the measurer at the speed v, the measured frequency would *decrease* by the factor indicated in equation (8). This would mean that an emitted frequency of radiation in the middle of the visible light spectrum would be seen to shift toward the red end. This is called the 'red shift'. On the other hand, if the source of radiation is moving toward the measurer, equation (8) would be altered by taking $v \rightarrow -v$, implying that the measured frequency of the radiation would be greater than the emitted frequency; in this case emitted radiation in the middle of the visible spectrum would be seen shifted toward the blue end of the spectrum. This is called the 'blue shift'. (The wave propagates toward the observer in a direction opposite to that of its source, hence the sign change in the numerators above, compared with equations 8.6.) Thus if one should see a source of radiation, say a star, emitting spectra that shift toward the red end, one would be able to deduce that this source is moving away from the observer. Similarly, if the observed radiation is 'blue shifted' this signifies that it is moving toward the observer. An alternating red shift and blue shift would indicate that the source of radiation is in some sort of oscillatory motion, moving toward the observer in part of its cycle and away from the observer in the other part of its cycle. This 'Doppler effect' predicted by relativity theory has been well confirmed by the experimental facts.

Note that if the speed of the source of radiation, v, is very small compared with the speed of light, c, then the Doppler effect (8) may be approximated by expanding the term in the square root (in a binomial expansion) maintaining the first two terms, to give:

$$(8') \qquad \nu' \simeq \nu(1 - v/c) \text{ Hertz}$$

This is the result obtained for the Doppler effect in classical physics, for example in the case of Doppler shifted sound waves (in which case, c would be the speed of sound). Thus, Einstein's theory of special relativity gives the same result as the classical Doppler effect (as an approximation), when v is very small compared with c. But in the general case, the Doppler formula (8) is much different; it is the latter that has been well confirmed in the observation of Doppler shifted electromagnetic radiation.

A further comparison with the classical predictions comes from the case in which one observes the radiation in a direction that is perpendicular to the direction of propagation. In this case, $\theta = 90°$, so that $\cos\theta = 0$ and equation (7) predicts the 'transverse Doppler effect':

$$(9) \qquad \nu' = \nu/[1 - (v/c)^2]^{1/2} \text{ Hertz}$$

This effect is not predicted at all in classical physics. Thus to observe it would be a strong confirmation of the theory of relativity's validity. Though it is difficult to see, because in an expansion its leading term in powers of (v/c) would be $(v/c)^2$ (a very small number for ordinary values of v), the transverse Doppler effect has also been verified in experimentation, thus making the validity of the theory of relativity even more firm.

Finally, it is important to note that if one should interchange the parameters for the 'observed' and the 'observer' in the expression for the Doppler effect (that is, if the emitter becomes the absorber and vice versa) then we must interchange

$$\nu' \leftrightarrow \nu \quad \text{and} \quad v \leftrightarrow -v$$

in equation (8)—which leaves equation (8) unaltered. This is not the case with the classical Doppler effect equation (8′). The reason for the asymmetry in classical physics, such as the Doppler effect for sound waves, is that there is a medium there that plays the role of conducting the propagating waves. For it follows in the classical case that if the source of the sound waves, say, would move from the listener, or if on the other hand the listener should move from the sound source, the corresponding Doppler effects would be different. In the latter case, equation (8′) gives the correct frequency shift. But if the source moves relative to the listener, the measured frequency would be:

$$\nu' = \nu/(1 + v/c) \tag{8''}$$

Expanding this in a binomial series, it takes the form:

$$\nu' = \nu[1 - v/c + (v/c)^2 - (v/c)^3 + \ldots]$$

Clearly, if v/c is a very small number and we could accurately neglect all of the terms in this series beyond the linear term, v/c, it would have the same form as equation (8′). But this is not exactly the case, and therefore this Doppler shift in sound, when the listener moves relative to the source of sound, is slightly different than the Doppler shift for the case of the source of sound moving relative to the listener. The reason for the difference is the existence of an *absolute medium* that is there to conduct the wave motion.

The *reason* for the Doppler effect in electromagnetic radiation is not the same as in the case of sound waves. In the former case, *there is no medium.* This was an important consequence of Einstein's special relativity theory and it was a result empirically verified in the Michelson-Morley experiment, though the latter experiment came before Einstein's theory (it was not intended as a test of the theory). Contrary to the case of sound waves, where a medium is involved, the Doppler effect as a consequence of relativity theory comes from the time transformations that must be imposed from one inertial frame to another to ensure the objectivity of the law (in the case of electromagnetism, the Maxwell field equations). It is simply that the electromagnetic force field changes its expression in transferring the description to different relatively moving reference frames. Thus, the frequency measure (in Hertz = cycles per *second*) is a frame-dependent quantity in relativity theory. It is a language change, according to special relativity; it is not a physical change due to the interaction of the wave with its physical medium as in the case of sound propagation. Nevertheless, one does measure this frequency change of electromagnetic radiation from a moving source. Similarly, if one looks from the air into a pool of water with a meter stick at the bottom, the stick appears to be less than a meter long.

But the meter stick has the same matter composing it whether it is at the bottom of the pool or in the air; the meter stick is physically the same stick whichever frame of reference it is observed from. Similarly, a source of radioactivity that may emit N decay products in its lifetime will not alter this number of decay products, whoever may be observing this decay process, from whatever reference frame—even though the radioactive source may seem to emit these N particles at a different rate as viewed from a reference frame that is in motion relative to it.

Light and Matter in General Relativity

If any material system is 'closed' according to general relativity theory, and if the atomistic model of matter is to be replaced by the continuous field model in accordance with Einstein's theory, then light must be interpreted differently than it is in the conventional view. This is a new view of light that would have to eliminate the 'photon' concept as an independent atom of light. Instead, 'light' would be not more than a connective relation between the interacting components (of matter) of a system. This is the relation that expresses their mutual electromagnetic coupling at a large distance. Since the very existence of matter implies the necessity of expressing its fundamental fields in a curved space-time, according to relativity theory in its most general form, it follows that in principle 'light' must be defined in terms of a propagating disturbance between material components of a closed system, described in a curved space-time.

With this view of a material system as closed, light is not a thing by itself, but rather it is like a ripple moving along a rubber band, connecting two blocks of wood. After one of the wooden ends has been 'flicked', the rubber connecting band will exhibit a vibrational mode of the entire system. The 'ripple' is not a thing by itself, it cannot be removed from the rubber band or the blocks of wood as a separated object. This analogy is not a precise one, since the concept of the 'closed system' that is without actual parts (it is truly *one*) is an abstract representation of the nature of matter. Its validity may only be checked by comparing the experimental facts with logical and mathematical deductions that follow from the closed system concept. It cannot be seen directly.

The important role of abstractions here should not worry the twentieth-century scientist. For we have learned that many truths in contemporary physics are not directly sensed by our bodily instruments or our artificially constructed measuring instruments. Rather, many of these truths must be deduced in logical fashion from a *hypothetico-deductive* method of investigation. It is the underlying hypotheses that we investigate; these play the role of the truth we seek, in the form of 'universals'.

Thus, the analogy between the phenomenon of light according to the logic of Einstein's general relativity and the rubber band model described above is close only in the similarity that both are in terms of connective relations—though one is abstract and the other is perceivable with the senses. Indeed, one may think that he *sees* light. But this is certainly an illusion. For what he sees is a response of the matter of his brain to other electrically charged matter. The observer then *speculates* that there is other matter that emits a signal of light that is then received by his eye, sending the signal to the brain that he calls 'light'. But this is not the only possible interpretation of this brain response. If it is true, as we see in general relativity theory, that light is a connective relation between electrically charged matter, and nothing else, then indeed, light is not an independent entity.

Thus the conceptual view of 'light', according to the theory of general relativity, is quite contrary to the model of what light is supposed to be in the quantum theory, as a collection of 'photons', each characterized by a wave-particle dualistic description. With general relativity, the 'particle' aspect of light is illusory; there is no actual dualism. Rather, light, in this theory, is a continuous field that *is* the coupling of the material components of a continuous material medium that is, in principle, *without parts*.

In this generalized description of electromagnetism in general relativity, the basic entity is then not a 'thing'. The 'elementary interaction' replaces the individual charged matter source as basic in the meaning of the field solutions of the laws of nature. With this view, the elementarity of interaction prescribes a different interpretation of the Maxwell field equations for electromagnetism. Here, they must be interpreted as *identities*. That is, particular rates of change of the electric and magnetic field variables that appear normally on the left-hand sides of these equations, are taken to be just another way of expressing the charge and current densities that appear on the right-hand sides of these equations. Thus, if there is no charged matter in a system (the vanishing of the right-hand sides of Maxwell's equations) then there can be no electric and magnetic fields to talk about. The reality of electromagnetism comes from the interactions between charged matter fields, expressed as the products of electromagnetic field variables for one domain of matter, coupled to the charge and current density variables of another domain of matter.

It then follows that the solutions of Maxwell's field equations in the case of no charged matter sources are unacceptable as physically meaningful. The latter fields, called the 'radiation solutions', are *out of context,* according to the meaning of these equations in general relativity. Now, the 'photon' is simply a quantum associated with these 'source-free' solutions of Maxwell's equations. Thus, 'photon' is an unacceptable concept according to the field theory of electromagnetism in general relativity. This is indeed a happy circumstance in the relativistic description of micromatter, for it is the photon field that has thus far caused most of the trouble in the attempts to formulate a finite, mathematically consistent quantum field theory ('quantum electrodynamics') that would be compatible with both the theory of special relativity and the rules of quantum mechanics.

In research that I have been engaged in since the early 1960s, I have claimed to show that on rigorous grounds, the 'photon' is indeed a superfluous concept in electrodynamics—that is, all of the empirical facts that are supposed to require the use of the photon concept in their explanations can be re-derived in a mathematically consistent way from a theory that is fully based on a continuous field theory of matter in general relativity without the need to introduce the photon concept.[81]

The original introduction of the photon by Einstein, in the early part of the twentieth century, was of course seminal for the quantum revolution that we have discussed previously (in chapters three through seven). It was the first suggestion of the notion of 'wave-particle dualism' carried forward to matter by de Broglie. The 'particle' aspect of the quantum of electromagnetic radiation, the 'photon', was successfully applied to several quantum phenomena in the microdomain, such as the photoelectric effect: the release of electrons bound in a metal to their conducting state when absorbing beams of light, in such a way that the electrical current depends linearly on the frequency of the absorbed light. It also explained the Compton effect: the scattering of electrons (in free space) by impinging single frequency photons, in such a way as to make it appear that indeed the photons are localized particles of light.

Recall that the successful explanations of these phenomena were, nevertheless, based on the proposal of a *quantized interaction* between the electromagnetic radiation field and matter. Now if the matter alone is quantized, that is, if the matter is only able to absorb discrete quantities of energy, angular momentum, and so forth from the radiation field, then this matter would act as a filter to any incoming radiation, whether or not the radiation itself is quantized. It may also be the case that, in order to correctly represent the data, the matter field variables could depend on the space and time variables in such a way as to predict a very rapid and *apparently* localized absorption of radiant energy (though not exactly discrete). That is to say, in this case, the absorption would be highly peaked in time, *though continuous*, rather than actually discrete. This view, then, does not necessitate the photon concept for the description of the coupling between micromatter that is electrically charged and an electromagnetic radiation field.

Delayed Action-at-a-Distance

The idea that most of the contemporary experiments that are supposed to entail photons could be explained within the context of electromagnetic theory without photons is not new. It corresponds to the idea of *delayed-action-at-a-distance* discussed by many authors since the 1920s, such as H. Tetrode, G.N. Lewis, Einstein, and J.A. Wheeler and R.P. Feynman.[82] Wheeler and Feynman's view introduces a system of *n* space-time trajectories of a set of mutually interacting, electrically charged bodies, where the mutual interactions appear symmetrically as follows: what appears from one frame of reference as an emitter of radiation and an absorber of radiation to react at some later time to the emitted signal, could equally appear from another reference frame in which the emitter is the absorber and the absorber is the emitter. In this case, each of the interacting material components of the coupled system is simultaneously emitting a signal to the other component—where the signal is not a free photon, but is, rather, a propagating force that is mutual (analogous to the setting up of a standing wave along a rubber band that connects two blocks of wood, discussed previously). In this case, also, there is no need for the free 'photon' to enter the picture at all.

And the symmetry of 'emitter' and 'absorber' is, in fact, required by the principle of relativity, as we have mentioned earlier. The idea here, then, is that we start at the outset with a closed system of matter components and no free radiation to deal with.

This description then entails a system of all particles and no mediating free fields. There would be, say, N particles, half being emitters and half absorbers. According to the theory of special relativity, we must impose a four-dimensional space-time for each of the trajectories; we are thus dealing with a four-N-dimensional space-time. The symmetry between the 'emitters' and the 'absorbers' implies that the solutions of Maxwell's equations must be represented with 'retarded' and 'advanced' solutions on an equal footing, corresponding to the 'future' and 'past' portions of the light cone, as discussed in chapter eight. The 'retarded' solutions usually refer to the sequence *cause → effect:* radiation is emitted *earlier* (cause) and it is absorbed *later* (effect). The 'advanced' solution is often referred to in terms of the backward sequence in time, *effect → cause*—which appears to violate the *law of causality*. But there isn't really such a violation in relativity theory. For if one should switch the names, 'emitter' and 'absorber', implying a switch of the words, 'cause' and 'effect' (corresponding respectively to 'earlier' and 'later') then no real change occurs in the overall description of the *mutual interaction* of the material components of the coupled system. In this view, what is elementary are not the separate things, 'emitter' and 'absorber'; instead it is the *elementary interaction* that is fundamental, an entity in principle without separable parts. As we have discussed earlier, this is the view that is in fact dictated by the philosophical approach of the theory of relativity.

One of the troubles that was recognized when the 'delayed-action-at-distance' view was pursued (in the latest attempt by Wheeler and Feynman), was that according to the Maxwell field theory some of the coupled radiation produced by the mutually accelerating charged particles escapes from the absorbers. It then became necessary for them to postulate the existence of infinite absorbers to surround the universe, in order to take up this excessive radiant energy. It was primarily for this reason that they abandoned the model of delayed-action-at-a-distance.

However, when one fully unifies the delayed-action-at-a-distance concept with the theory of relativity (that logically requires it) one has a model of matter different from that of Wheeler and Feynman, that does not suffer from these drawbacks. For in this case the basic variables of matter are not the trajectories of singular, charged particles. They are rather the coupled, continuously distributed fields of the matter components of the system, all mapped in the same four-dimensional space-time. With the mathematical expression of this view, one has a highly *nonlinear* field theory for the closed system, in contrast with the Wheeler-Feynman many-particle model, and in this field view there is no radiation escaping into the universe and therefore no need to introduce an infinite absorber to surround the universe. With this view, then, the 'photon' (a quantum of the electromagnetic 'free field') disappears from view as a superfluous concept.[83]

Pair Annihilation and Blackbody Radiation

From the conceptual point of view, then, general relativity rules out the 'photon concept'. There are, however, still two sets of experiments that are supposed to entail photons when there is no charged matter present. One of these provides the data that originally led Planck to the photon concept (in the context of the 'old quantum theory')—the experiments on blackbody radiation (discussed in chapter three). The second supplies the data conventionally interpreted as 'pair annihilation'. The former experiments were interpreted by Einstein in terms of a box full of 'photons' that are in thermodynamic equilibrium with the walls of the container, 'photons' whose statistics make them appear as though they are independent of any matter. The latter was revealed in the fact that the blackbody radiation curves were all the same for different material boxes, and are thus not a manifestation of the charged matter of the walls of the box, or so it seems. The experiments on 'pair annihilation' seem to reveal the disappearance of tracks of two (oppositely charged, equally massive) particles, such as an electron and positron, as they come near each other. This 'disappearance' is then correlated with the 'creation' of radiation in the form of two photons that move away from the scene of the pair annihilation, in opposite directions with opposite angular momenta. Thus, the supposedly created radiation exists at times when the matter has disappeared, according to this interpretation of these data.

In my research program, I have discovered that there are exact solutions for the coupled, nonlinear matter field equations for a particle and antiparticle (such as an electron and positron or a proton and antiproton, etc.) that explain both of these phenomena (blackbody radiation and 'pair annihilation') without the need to introduce the photon concept or actually 'annihilating' matter of any sort.[84] The explanation came from a *bound state* solution for the particle-antiparticle pair. Summarizing the features of this solution it was found, first, that the total energy of the pair, when in this bound state, was exactly equal to zero (as was the momentum relative to any observer). This means that when the particle and antiparticle are (almost) free of each other, the energy is (almost) equal to $2mc^2$, whereas the energy in the bound state is zero. It was also found with this exact solution that the total angular momentum of the pair in this state is zero: it is an S-state.

With the standard view of positronium (the bound electron-positron) the ground state is supposed to be just a few electron-volts below the free state where the energy is $2mc^2$. But with this new solution, the true ground state is $2mc^2$ (which is of the order of a million electron-volts for the electron/positron pair, or about two billion electron-volts for the proton-antiproton pair) below the free state, when the particle and antiparticle are free of each other. Therefore, in the actual ground state of the pair, corresponding to zero energy, ($2mc^2 - 2mc^2$), the particle and antiparticle are so tightly bound that they do not readily give up energy and momentum to their surroundings. Thus they appear to be invisible. Recall that one does not actually *see* particles, say in a cloud chamber. One rather sees the tracks that they create by giving energy and momentum to their surroundings to create these tracks. Thus, if they did not give up such energy and momentum because of the particle and antiparticle going into a deeply bound state, one would no longer see them, but they would still be there. Indeed, if this tightly bound pair should be 'ionized' by supplying them with energy that is the order of $2mc^2$, they would make themselves 'visible' by creating tracks once again. The latter is an experimental fact; the seeming appearance of particle and antiparticle tracks out of a vacuum is conventionally interpreted as 'pair creation'. But with the model we have investigated within the domain of the general relativity approach to matter, there is no real annihilation or creation of matter. It only appears to be so because of the disappearance of tracks in a cloud chamber or their re-appearance somewhere else at some other time, under the proper physical circumstances.

The solution of the matter field equations for the pair also led to the dynamics of this pair in the 'ground state' as corresponding to a charged current field. This field would be such that two measuring apparatuses on either side of the plane of polarization of the charged currents in this state would respond to oppositely polarized currents (by direct electromagnetic coupling to their corresponding electric fields), correlated with a 90° phase difference, polarized in the plane that is perpendicular to the direction of propagation of the interaction with the measuring devices.

These predictions are in exact correspondence with the data on pair annihilation that are conventionally interpreted in terms of the actual annihilation of matter with the simultaneous creation of two photons of this sort (correlated in their polarizations as described above). Thus, to explain these data there is no need to introduce any photons nor to assert that charged matter is in fact 'annihilated'.

With this result for the single electron-positron pair (or proton-antiproton pair, and so forth) it may be conjectured that there is no reason why any spatial domain could not be populated with some arbitrary number of such pairs, each in their ground states of null energy and null momentum and angular momentum. This would be the 'physical vacuum' according to this field theory, consisting of a *countable* number (though invisible set) of pairs. It was found in this research program that such a background 'physical vacuum' can account for the observations associated with the evidence of the spectrum of blackbody radiation that was studied by Planck and later by Einstein (though here again without the need of photons).

With the derived features of the dynamics of each of the constituent pairs of the physical vacuum of this theory, and with the assumption that the detecting apparatuses respond to *distinguishable* matter fields of the gas of such pairs in a cavity, at some equilibrium temperature, one may use the classical Boltzmann-Maxwell statistics (rather than the Bose-Einstein statistics conventionally evoked for the statistics of a 'photon gas'), to determine the averaged properties of this system of matter fields in the box. The mathematical problem posed by this situation is exactly the same as that of Planck, who also used Boltzmann-Maxwell statistics for his model of distinguishable vibrations of the electromagnetic radiation field in the cavity, at fixed temperature. It is interesting to note, aside from the theory proposed here, that Planck's analysis of blackbody radiation that used Boltzmann-Maxwell statistics, and the quantum view of indistinguishable 'photons' that used Bose-Einstein statistics, both yielded the same Planck formula for the spectral distribution of blackbody radiation! Thus, the quantum agreement with the data on blackbody radiation was not a crucial test of quantum mechanics.

Because frequency dependence of the matter field solutions for the pairs in this field theory is that their energies, when they are in their ground states of null energy-momentum, are linearly proportional to their frequencies (as was found also by Planck in the case of electromagnetic vibrations in the cavity), this theory predicts the same spectral distribution for blackbody radiation that was found by Planck, in agreement with the data on this effect, discussed in chapter three.

Summing up, a rigorous derivation of an exact solution of the coupled matter field equations for any particle-antiparticle pair was found to be in strict accord, both mathematically and conceptually, with the requirements of the theory of relativity. This duplicates in its predictions all of the experimental facts having to do with i) pair annihilation and creation (though without actually annihilating or creating matter) and ii) blackbody radiation (both effects without the need to introduce 'photons'). Along with these predictions, all other experimental data that are supposed to entail photons when matter is also present may be explained with this field theory without the photon concept. Thus, one may conclude that within the context of this field theory it is possible to dispense with the photon concept altogether.

The Generalized Mach Principle

Finally, if one should impose the idea of the *Mach principle,* applied to all aspects of interacting matter, rather than only to its inertial manifestation, then the interactions that appear in the generally relativistic field theory are automatically prescribed. This is the third axiom that must be included, in addition to *the principle of relativity* and the *principle of correspondence,* to form a mathematically and logically consistent theory of matter. The *generalized Mach principle* asserts that all of the features of matter previously considered intrinsic from the elementary particle point of view (electric charge, magnetic moment, . . . , in addition to mass) must now be considered as measures of *intrinsic coupling,* that is, expressing the elementarity of interaction, rather then the elementarity of particles, *of* (rather than *in*) an entire, single, continuously closed system that is, in principle, without parts.[85]

The requirement of the *generalized Mach principle* then does away with all remnants of the atomistic model in the fundamental explanation of matter. Further, it forces the fundamental theory of matter into a definite form, in terms of nonlinear, coupled matter field equations. It yields the inertial feature of matter in terms of field properties that depend on the curvature of space-time. Thus, all the matter of a closed system does, in principle, yield the inertial mass of any 'test body' of that system, since the curvature of space-time is indeed a manifestation of all of the matter of the system. If all of the matter of a system would then be continuously depleted, the mass of the test body should correspondingly diminish, until the mass would vanish in the (unreachable, ideal) limit of a perfect vacuum (except for the test body itself). In this limit, according to the generalized Mach principle, *all other* manifestations of the test body would also automatically vanish, such as its electromagnetic properties (electric charge, magnetic moment, and so forth). With the *generalized Mach principle,* the *principle of relativity* and the *principle of correspondence,* it was shown in this research program that the field equations that are an explicit representation of the inertial manifestation of matter have, as a low energy, nonrelativistic limit, the formal expression of quantum mechanics. That is, in this theory, the formal expression of the probability calculus that is 'quantum mechanics', is really a low energy approximation for a field theory of the inertia of matter. The latter field theory does not, generally, have any of the features that are required by the philosophy of quantum mechanics. It is not expressible according to the rules of a probability theory, it is nonlinear and it does not relate in principle to the elementarity of particles of matter. Still, it gives back the formal expression of quantum mechanics, where the latter theory works: in the low energy description of matter in the domain of micromatter.[86]

Summing up: the logical basis of the theory of relativity, in its most general form, is dichotomous with the logical basis of quantum mechanics; they are competing theories of elementary matter. When it is fully exploited, the theory of relativity implies that matter must be fundamentally characterized by total objectivity, continuity, determinism and nonlinearity. On the other hand, the quantum mechanical view of matter, in fundamental terms, is based on the ideas of subjectivity, atomism, nondeterminism and linearity. These philosophical differences in the fundamental bases of these two approaches to matter will be discussed in detail in the next chapter.

Ten

Fundamental Conflicts Between the Quantum and Relativity Theories

The science of physics—a study toward a fundamental explanation for the behavior of inanimate matter in terms of underlying principles—is now at a very exciting and challenging stage, perhaps more so than at any previous time in the history of physics. On the one hand, there is presently more accumulated information and ongoing activity in both experimental and theoretical physics than at any previous time. But on the other hand, little has been achieved in regard to our *fundamental understanding* of matter beyond the stage that was reached in the 1920s, about 60 years ago, when the ideas of the two major scientific revolutions of the twentieth century had crystallized: the quantum theory and the theory of relativity, both considered (by their chief spokesmen) as fundamental theories of matter.

It is my purpose in this chapter to demonstrate a conviction that any real progress in our understanding of matter can only come if we agree to reject some of the currently held notions of *either* the quantum theory *or* the theory of relativity. Further progress might then be achieved by properly generalizing one of these approaches or the other, as relating to a basic explanation of the nature of elementary matter.

It will be demonstrated in this chapter that to achieve the goal of a logically consistent, sufficiently complete theory of matter, it must be recognized that when the quantum mechanical theory of elementary matter and the theory of relativity are each fully exploited in terms of their respective conceptual and mathematical implications, they cannot peacefully coexist. This is because some of the axioms that underlie one of these theories are not logically compatible with some of the axioms that underlie the other theory. If the physics of the future is to proceed from the conceptual and mathematical basis of one of these theories of elementary matter, then it must necessarily reject the other. *The problem of matter* is then to choose one of these theories as a basis for future progress in physics, while incorporating the successful mathematical features of the rejected theory.

Atomism versus Continuity—An Outstanding Conflict

In attempting to make judgements about possible paths of inquiry that could lead to resolutions of fundamental problems in physics, I believe it could be useful to make a close study of the themes that persist throughout different periods in the history of physics. One theme that has been recurrent through the ages deals with the question of whether matter is fundamentally discrete or continuous. Something happened in the contemporary period, relating to this question, that has never happened before in the history of physics. In the past only one revolution in science happened at a time. But in the twentieth century two scientific revolutions occurred almost simultaneously. These are the quantum theory in its new form, expressed primarily by Bohr and Heisenberg, and the theory of relativity in its new form: that of the theory of general relativity, as expressed primarily by Einstein.

The theory of relativity, when fully expressed in terms of its axiomatic basis, indicates a continuum view of elementary matter (in terms of the continuous field concept) while the quantum theory in its new form of quantum mechanics, is essentially an atomistic view of matter, though of a different sort than the classical atomistic approach, as we will discuss below.

The compelling point about the *simultaneous* occurrence of these two scientific revolutions is that when their axiomatic bases are examined *together,* as the basis of a more general theory that could encompass explanations of phenomena that require conditions imposed by both theories of matter (such as current 'high energy physics'), it is found that the widened basis, which is called a 'relativistic quantum field theory', is indeed logically inconsistent because there appear, under a single umbrella, assertions that logically exclude each other.

In the next section I will compare the full set of concepts that are implied by both of these theories, indicating in detail their compatibilities and incompatibilities. In the subsequent sections I will discuss the reasons for the necessary requirement that these two theories be unified, based on their own terms. I will then show that indeed the quantum and relativity theories cannot be unified, for explicit logical and mathematical reasons. Perhaps, then, this implies a third revolution in physics. It will be my contention that such a basic change will not be abrupt; rather it will evolve out of only one of the existing theories, while maintaining the mathematical form of the rejected theory under the limiting conditions where it has been empirically successful. I will then present my reasons for favoring the theory of relativity, indicating that such an advance would view the concept of 'wave-particle dualism' in the history of physics as a necessary intermediate stage between 'particle monism' and 'wave monism', that is, as bridging an atomistic and a continuum approach to elementary matter.

Conflicting Concepts of the Quantum and Relativity Theories

A comparison of the conflicting concepts of the quantum and relativity theories, when fully exploited, is shown explicitly in table 1. These will now be discussed in order of their appearance there.

Table 1

SOME OPPOSING CONCEPTS OF THE QUANTUM AND RELATIVITY THEORIES

Quantum Mechanics	*Relativity Theory*
A. Principle of Complementarity, pluralism	Principle of Relativity, monism

B. Atomism: open system of separable things; elementarity of 'particle'.	Continuity: closed system of distinguishable manifestations; no parts; elementarity of interaction.
C. Logical positivism	Abstract realism
D. Subjective → essential features of matter depend on measurement by macroobserver on micromatter, not vice versa. Essential role of probability at explanatory level. Non-causal relation between observer and observed, asymmetrically. Macrovariables from classical physics, microvariables from quantum physics.	Objective → features of matter are independent of measurement. No difference between micro and macromatter. Probability descriptive, not explanatory, Symmetry between observer and observed. All variables from the same (covariant) laws—all obey the same rules.
E. Nondeterministic → properties of matter not predetermined.	Deterministic → all matter field variables are predetermined.
F. Linear, homogeneous eigenfunction differential equations. Linear superposition principle. Special reference frame for the measuring apparatus. Separate space-time for each particle component of a system.	Nonlinear, nonhomogeneous, integro-differential equations. No linear superposition principle. No special reference frame (covariance = total objectivity). One four-dimensional space-time for the mapping of all field components of a closed system.

A. In my view, the most significant starting point of the philosophical view of the quantum theory is Bohr's *principle of complementarity*. This concept grew out of the idea of wave-particle dualism in physics, though it may have had earlier thematic origins in philosophy. It was Einstein who first suggested this dualism, applied to the seemingly dualistic nature of electromagnetic radiation in the microdomain, characterized in terms of the quantized units of this field, called 'photons'. After de Broglie successfully extended this dualism to matter, as in the description of electrons, his speculation was confirmed in electron diffraction experiments, as we have discussed earlier in the text.

Bohr then asserted the generality, in principle, of incorporating *opposing* bases that may be separately true at different times. He thus proposed the idea that seemingly logically exclusive propositions can *both* be true as complementary aspects of a general description of the true behavior of radiation or matter, in the microdomain. Such a philosophy is then 'pluralistic', whereby one assumes that several simultaneous levels of explanation exist, based on the conditions of observation in experimentation, even though when considered together these concepts are logically dichotomous. This is *Bohr's principle of complementarity.*[87]

In opposition to this pluralistic view, the theory of relativity starts with an assertion that the laws of nature must be purely objective (this is *the principle of relativity*) thus claiming that the expressions of the laws of nature are independent of the reference frames that any observer may use to express these relations (relative to his own reference frame). With this fundamental approach, 'observing' the effects of radiation, for example as 'particles' or as 'waves', under correspondingly different sorts of experimental conditions, must be represented in terms of comparisons of these respective phenomena from different possible frames for a single substantive system. That is, it may *appear* to be a wave phenomenon from one reference frame (experimental set-up) or it may *appear* as a particle phenomenon from another, but these *appearances* are in fact based on a single underlying logically consistent law, a law based on a set of logically consistent propositions.

Implied in the relativistic approach to a theory of matter and radiation is the idea that there is a single underlying order, as expressed in terms of fully objective laws of nature. That is, it is assumed that underlying the behavior of elementary matter (in any domain, from elementary particle physics to cosmology) there must exist a single, logically ordered universe. There can be no conceptual lines of demarcation between one set of axioms, to underlie one sort of physical phenomenon, and another set of axioms that underlie other sorts of physical phenomena. For example, in theories of matter, the conceptual notion of 'wave', which entails continuity and rules of combination that include 'interference', is logically exclusive from the notion of 'particle', which in contrast entails locality, discreteness and the rules of ordinary arithmetic for their combination. If there is to be a single conceptual basis for the laws of matter, it cannot then include the logically dichotomous concepts of 'particle' *and* 'wave' in fundamental terms. It then follows that *Bohr's principle of complementarity* is automatically rejected by the holistic approach of relativity theory, and *vice versa*—the monistic approach of relativity is automatically rejected by the dualism of complementarity. In relativity, then, it must be assumed that there is a single explanatory level for the workings of the universe, in any of its manifestations, from the small to the large. Such a view is philosophically monistic.

B. With the quantum mechanical view of elementary matter, the world is supposedly atomistic. As in Newton's classical approach to matter, the universe is assumed at the outset to be the sum of its parts. These are entities that by definition may be separated from the whole without in any way altering the basic characteristics of the whole. On the other hand, when exploiting the underlying symmetry requirement of the theory of relativity, it must be assumed that the universe, in any of its domains, is basically represented by a continuum for a closed system—that is, a system that is truly without separable parts. This conclusion follows from the principle of relativity, which requires the laws of nature to be totally objective (covariant) with respect to *continuous* (and continuously differentiable) transformations from the space-time language of any given reference frame to the space-time languages of any other reference frame in which one may wish to compare the forms of the laws of nature.

Thus we see that the elements of matter that are supposed to be fundamental, according to the quantum mechanical view, are distinguishable 'parts', called 'elementary particles', while the fundamental concepts of the continuum representation of matter, according to the theory of relativity, are its (infinite variety of) distinguishable manifestations (modes) of a continuum. This may be characterized by the *elementarity of interaction*. Metaphorically, it is similar to the distinguishable manifestations of a continuous pond in terms of its 'ripples'. These are modes of the pond, they are not parts in it. There may be a transfer of energy and momentum between the ripples of the pond, yet, they are not separable, distinguishable entities on their own: one cannot remove a ripple from a pond as a separate thing. Nor are the ripples precisely localizable, except insofar as where they might peak at one time or another.

In this regard, the full exploitation of the theory of relativity in the problem of elementary matter implies that in fact there are no free, 'localizable', separable particles of matter. The recent interesting research activities concerned with 'Bell's inequalities' indeed highlight the rejection by the formal structure of quantum mechanics of the notion of localizable particles of matter.[88] But the present day interpretation of these experimental results, that seem to violate Bell's inequalities, thereby upholding the Copenhagen view, is certainly not unique. For the continuous field concept that is implied by the theory of general relativity, as a fundamental theory of matter, is also compatible with the same experimental facts. Yet, as I will demonstrate in this chapter, the theory of relativity is incompatible, from the logical and mathematical views, with quantum mechanics. Quantum mechanics would then be expected to serve as not more than an approximation for the mathematical form of the exact relativistic theory of matter, in the low energy limit. As we have discussed in previous chapters, this would then be a 'linear approximation' for a nonlinear field theory of matter.

C. The epistemological approach of quantum mechanics, as we have indicated earlier, is that of *logical positivism*. This is an approach to knowledge based on the 'principle of verifiability':

the assertion that the only 'meaningful' statements in science, whether expressed mathematically or in ordinary language, must be empirically verifiable.

We see that the notion of 'wave-particle dualism' and, generally, the principle of complementarity, are ideas that are consistent with the philosophical approach of logical positivism. This is because with the latter views one may assert the *truth* of logically opposing models, so long as at the different times when different sorts of observations are made in experimentation they reveal consistency with these separate, apparently conflicting views. With this philosophy, one needn't say that the 'particle' picture is *the* true one, that is, that the particle model *underlies* its wave aspects, or vice versa. Rather, at the different times when one observes the electron as a particle, it *is* a particle and when one observes the electron as a wave, it *is* a wave. According to this epistemological point of view, this is all that can be said in a meaningful way about the electron.

On the other hand, the theory of relativity as a theory of matter is based on an epistemology of *abstract realism*. This is an approach to knowledge whereby it is assumed that there is a real world, independent of whether or not anyone may be perceiving it in one way or another. The explanation of the real world is then in terms of *underlying principles*. What we do see, as observers of phenomena, is then a sort of 'projection' of this underlying reality. The adjective, 'abstract' is meant to convey the idea that we do not generally observe the truths of the world, directly, but rather we must deduce them from an interpretation that is lodged in some universal theory. That is, the claim is that we arrive at scientific truths by means of the hypothetico-deductive method, using our powers of reason as well as observation, to reach particulars (the predictions of a theory) from a universal (the 'unobserved' hypotheses that are supposed to underlie the phenomena). It might be argued that here there is no prescription for arriving at the *alleged* universals at the outset, that is, the claimed laws that are to underlie the observable phenomena. What we must answer is that we arrive at these universals mostly from the *hints* that are received from the observable facts of nature, *though also from our imaginations.* It is important to recognize that the hints from the phenomena could very well be false, and we should be ready at any time to abandon a particular point of view on the underlying explanation for a particular physical phenomenon if it does not withstand the scrutiny of the experimental facts and the theoretical requirement of logical consistency. In any case, one must always rely on the experimental facts to confirm (or refute) the claim of a theory that it is a 'law of nature'.

D. The approach of quantum mechanics entails an *irreducible subjective element* in its conceptual base. In contrast, the theory of relativity, when it is fully exploited, is based on a *totally objective view.* This very important difference in the philosophies of the quantum and relativity theories has to do with the fundamental incorporation of the macroscopic measuring apparatus with the observed microscopic matter, *asymmetrically,* in the very explanation of the micromatter itself, according to the Copenhagen view, contrasted with the fully objective approach of relativity theory. The claim of the Copenhagen school is that there is no causal description of the coupling of the macroapparatus and the micromatter that it observes, thus there can be no certainty in the prediction of the outcome of a measurement of any of the properties of micromatter, *in principle.* It is further contended that the outcome of the (alleged non-causal) coupling of the macroapparatus to the micromatter are the only meaningful statements that can possibly be made about the matter.

It then follows from this Copenhagen view that at the basis of the physical laws of elementary matter there must be an *irreducible* probability calculus, a set of rules of probabilities that are the limiting form of the law of nature that pertains to elementary matter. It is further asserted with this approach that the physics of the macroapparatus must be the classical rules of Newton while the physics of the micromatter are the rules of quantum mechanics, that is, rules that obey the properties of the complex number probability amplitudes (mathematically described as the elements of a 'Hilbert space' of functions, described in chapter six). Thus we see that in the view of quantum mechanics, one must start at the outset with the coupled system (macroapparatus|micromatter), distinguishing each of these labels by the location of the line of demarcation between them. There is no strict rule about the location of this line, aside from orders of magnitude. But once it is located, to describe a particular experimental arrangement, the 'apparatus' is thereby *defined* for this experiment, represented mathematically according to the rules of classical physics. The remainder of the system (the 'micromatter') is then described in terms of basic properties that are in accordance with this arrangement, and the ensuing statistical description in terms of the rules of quantum mechanics.

However, moving the line of demarcation *arbitrarily* (so long as the quantities of action involved on the 'micromatter' side are comparable with Planck's constant h) the nature of the observer correspondingly changes and the predictions about the 'observed matter' also change. Since, in this view, the basic properties of elementary matter are taken to depend in part on the nature of the 'observer', that is, the choice of precisely where to put the line of demarcation between the observer and the observed, it must be concluded that quantum mechanics, as a fundamental theory of elementary matter, entails an irreducible aspect of subjectivity in its very definition of the elements of matter. This conclusion is, of course, fully consistent with the epistemological approach of logical positivism, discussed above.

In contrast, *the principle of relativity* requires at the outset that the laws of elementary matter must be independent of the reference frame in which they are represented, be this the frame of the 'observer' or that of the 'observed' (corresponding respectively to the 'subject' and the 'object'). Since relativity theory requires at the outset a symmetry in the laws of nature with respect to changing reference frames, say between the 'observer' and the 'observed' of the subject-object interaction relation, it follows logically that 1) the laws of matter must be in terms of an *entirely objective description,* and 2) the variables that relate to the 'subject' in a particular experimental set-up, and those that relate to the 'object' in this arrangement, must both be subject to the symmetry requirements of relativity theory, and thus they must both be described by the same sort of language. That is, there should be no difference between microvariables and macrovariables according to the theory of relativity as a fundamental theory of matter.

It also follows in the theory of relativity, as a fundamental theory of matter, that, in contrast with the quantum theory, probabilities do not play any fundamental role. The probability function can be useful as a 'tool', that the inquirer may wish to utilize whenever he or she cannot determine the complete set of variables that are assumed to underlie the physical interactions. But when this is done, the inquirer is still fully aware that there does exist a more complete description that *underlies* his or her probings. This is in the mode of thinking of Boltzmann's view of the role of statistics in describing a gas, in which the complete underlying theory of its behavior is in Newtonian dynamics for all of the constituent atoms of the gas. But in contrast, the quantum theory asserts that the probability calculus it uses is at the limit of all possible knowledge about the material world of micromatter; that is, quantum mechanics is asserted to be as complete a description as there possibly can be. This view then claims that the fundamental laws of nature are laws of chance.

E. The Copenhagen interpretation of quantum mechanics is *nondeterministic.* That is to say, the fundamental assertion is made that the trajectories of the elementary units of matter are *not predetermined,* independent of measurement, as they would be in the classical, deterministic theories. With the quantum view, it is then postulated that when one makes an observation of some physical property of micromatter, (*necessarily* using a macroapparatus) then a group of possible states of the micromatter may be 'projected out', weighting some of them more than others. The assertion is that the more accurately that one attempts to ascertain a particular physical property of this matter, with particular types of measurements for this type of information, the less accurately can some ('canonically conjugate') physical property of this matter be simultaneously specified. This is the *Heisenberg uncertainty principle,* discussed in more detail in chapter six. According to this theory of elementary matter, the actual states of the micromatter that the macroapparatus responds to must be expressed in the form of complex number 'probability amplitudes'. Thus, the linear superposition of more than one such probability amplitude, because they are complex functions, yields an interference pattern in the measured properties of the matter, no matter how rarefied this matter may be (even if it seems that one is viewing one 'particle' at a time). According to this theory, it is impossible to reduce the interference to zero because the measuring apparatus alone, in its coupling to *any* quantity of elementary matter, automatically generates a superposition of the states of this micromatter whenever a measurement is carried out. Such interference of 'probability waves' was the interpretation that the Copenhagen school gave for the wave aspect of particle behavior.

In contrast, the theory of relativity, as a theory of matter, implies that the laws of matter are totally *deterministic*. That is to say, the logical implication of this theory is that the laws of matter are fully predetermined, independent of measurements. But the determinism of general relativity theory as a theory of matter is not in terms of predetermined trajectories of the 'particles' of matter, as it is in the Newtonian view. It is rather in terms of predetermined continuous fields, all mapped in the same space-time. That is to say, the 'determinism' in the sense of a relativistic field theory of matter does not single out the trajectories of singular particles, deterministically parameterized with a time measure for each of them. In the relativistic field theory, 'determinism' refers to all aspects of a closed system that are coupled in a totally objective way, independent of any measurements that may or may not be carried out. It is the full coupling of the closed system that is predetermined, though not only in terms of a 'time' measure, but rather in terms of a full set of logical and mathematical (geometrical, algebraic and topological) relations between the interdependent field components that underlie its basic description.

F. Because of the conceptual basis of quantum mechanics, interpreted in terms of a particular sort of probability theory, this theory of matter implies that the laws of elementary particles of matter must obey a mathematical feature that is the *principle of linear superposition.* This principle implies that any solution of the basic laws of elementary matter may generally be expressed as a linear sum of any other of its solutions.

It is further asserted in this theory that generally the equations satisfied by the probability amplitude solutions must be *homogeneous* in these solutions. This means that the equations must generally have the form in which the solution appears explicitly on both sides of the equality sign. Together with the assumption of linearity, the basic equations of this theory of matter must have the *eigenfunction form* (discussed in chapter six):

$$\hat{O}\psi_n = o_n\psi_n$$

Recall our discussion in chapter six, stating that the interpretation of this equation in quantum mechanics is that the 'linear operator', $\hat{O}$, represents the 'act' of making a measurement (carried out by a macroapparatus) on micromatter, that is in the state ψ_n. The linearity of this operator is a consequence of the assumption that the observed micromatter has no dynamical effect on this apparatus. For if it did, the operator $\hat{O}$ would have to depend in some way on the state function ψ_n itself, and then the equation would be nonlinear, that is, the left side of this equation would then depend on powers of the solution ψ_n to higher powers than unity. The measured *n*th value of the property described in this equation, with the eigenfunction ψ_n, is denoted by the number (eigenvalue), o_n.

As discussed above, one can never 'see' the micromatter in the pure state, ψ_n, because of the interference that is *necessarily* introduced by the measuring apparatus, when it couples to the micromatter. Nevertheless, one may design an experiment that can come arbitrarily close to observing properties of this micromatter in a pure state. That is, one may be able to *approach* infinitely great resolution in a measurement of a particular physical property, but *in principle* the theory says that it is impossible to reach this limit exactly. What this means, if one should follow this Copenhagen view to its logical extreme, is that it would be meaningless to even talk about a system in a pure state, since the latter is only an ideal case that is independent of making any real measurement.

It is interesting to recall Schrödinger's starting point for his formulation of wave mechanics: his wave form of quantum mechanics. This was a nonlinear equation, the 'Hamilton-Jacobi equation' of classical mechanics. The solutions of this equation are the values of the mechanical action of the system. Schrödinger's 'quantization' was accomplished by converting each term of this nonlinear differential equation into a 'linear operator', then allowing the resulting 'operator equation' to 'act on' imposed state functions of the matter, then interpreting this as the 'act of measuring' a property of that matter.

To incorporate the coupling of matter and the radiation that it transfers in an interaction process, one then proceeds to 'quantum field theory' by converting (in an analogous fashion) the Schrödinger wave functions, ψ_n, into 'linear operators' themselves, letting them 'act on' a second space of state functions. This mathematical technique is called 'second quantization'. In this way, the 'nonlinear' interactions in the 'first quantized' Schrödinger formulation, that couples matter to radiation, become operators that then act on new state functions in a linear 'function space'. The new state function operators then represent the act of 'creating' and 'annihilating' different numbers (and sorts) of elementary particles. In this way, the principle of linear superposition is restored, allowing once again the use of the probability calculus prescribed by quantum mechanics.

In contrast with the fundamental linearity of quantum mechanics, the elementarity of interaction in the theory of relativity prescribes that the expressions of the laws of nature that are to underlie any material system must necessarily be nonlinear. This is because of the basic description of any material system as *closed,* without separable parts. With this type of formal structure of the field equations for matter, *in principle* nonlinear, there is no possibility of expressing their solutions as the sum of other solutions of the same equations. Thus, the type of probability calculus that is used in the quantum theory to represent elementary matter in fundamental terms is automatically excluded in this approach. Also, because the system of matter, according to a basic underlying relativistic theory, is closed, it would have to be said that if any constituent of the entire system were excluded at the outset, say for the purpose of mathematical approximation, then the general solutions for the 'reduced' system could not be obtained by a simple subtraction from the original mathematical description. In principle, one would have to start all over again to consider the new (reduced) system on its own, to predict its physical consequences.

Summing up, I have attempted to demonstrate in this section that the full conceptual structures of the quantum theory and the theory of relativity, as competing theories of elementary matter, are not logically compatible. Thus, to accept the basis of one of these theories it would be necessary to reject the basis of the other theory, if we are to formulate a logically consistent theory of matter. On the other hand, there are domains of physical phenomena where the conditions that evoke each of these theories overlap, such as present day high energy elementary particle physics. This is indeed the major dilemma of modern physics. It must be resolved before we can make any bona fide progress in our fundamental understanding of matter.

In the next section I will discuss the important explicit reasons for the *necessary* conditions for unifying quantum theory and the theory of relativity, leading toward a 'relativistic quantum field theory', and why it is that, thus far, no such theory has been formulated in a demonstrably mathematically consistent way.

Is the 'Quantum Jump' Compatible with the Theory of Relativity?

One of the outstanding reasons for the *logical necessity* of expressing the formal description of quantum mecahnics in a way that would be compatible with the requirements of the theory of relativity, that is, so that the equations of quantum mechanics would be in one-to-one correspondence in all possible inertial frames of reference (this is the minimal requirement, corresponding to special relativity) is that the complete process whereby one is made aware of the existence of micromatter entails both matter and radiation (that is, the 'signal' that is transferred between the material emitter and absorber components in the measurement process). If the rules of quantization are to be scientifically valid, they must apply to all of the microscopic components of a described system—the matter *and* the quantized radiation that is created when a 'quantum jump' takes place, when the emitter 'jumps' down from a higher energy level to a lower one.

The difficulty is the following: while the micromatter components of the system, called emitter and absorber, have a limiting nonrelativistic description (Schrödinger's nonrelativistic wave mechanics or Heisenberg's nonrelativistic matrix mechanics), the signal that partakes in the interaction (the transferred radiation) has no nonrelativistic limit. This is true if the interaction is electromagnetic, and the signal is then a 'photon'. (However, with slight modification the argument below may be extended to all of the other types of interaction, nuclear, weak, and so forth.) In 'particle language', the fact that the photon can only move at the speed of light, in any reference frame, can be expressed by the fact that the photon has no inertial mass. Thus an external force cannot cause the photon to speed up or to slow down, as one could do to a massive particle, such as the electron. All that can be done to an existing photon is to annihilate it by absorbing it in matter (as that matter simultaneously undergoes a 'quantum jump' of increased energy).

Since the signal component of the triad, *emitter-signal-absorber,* must necessarily be represented at the outset in a relativistic manner, the entire *triad,* that is a fundamental coupling of matter to radiation, must be represented at the outset with a theory that obeys the rules of quantum mechanics and relativity, simultaneously. The main reason that the triad, *emitter-signal-absorber* is indeed unbreakable in this theory is that the measurement is the elementary entity that we start from, according to the Copenhagen school. Thus, we must start in our theoretical structure from an *element of measurement,* which is *minimally,* the triad that entails an element of matter of the *observed* (the emitter) an element of matter of the *observer* (the absorber) and the radiation that is transmitted between them (the signal).

Once a theory of matter and radiation is structured, so that it can be compatible with the rules of both the quantum theory and the theory of relativity (this is called 'relativistic quantum field theory') one could take the nonrelativistic limit of the parts of the system that may be associated with inertial matter—the emitter and absorber—then hoping to recover the form of nonrelativistic quantum mechanics. But it is salient that it is *logically necessary* to formulate the theory at the outset in terms of a mathematically consistent relativistic quantum field theory, that rigorously incorporates the 'quantum jump' of the matter in the microdomain and the radiation transferred between emitter and absorber.

The well known trouble that was found, when the first attempt was made to structure a relativistic quantum field theory (in the 1920s) that would simultaneously obey the rules of the quantum and relativity theories, was that the resulting equations had no solutions. But it is the set of solutions of *these* equations that are supposed to rigorously relate to the data on the phenomena of elementary matter. Thus, with no solutions it had to be admitted that quantum mechanics failed at the outset, in the 1920s, when it was formulated. For the part of quantum mechanics that was empirically successful, nonrelativistic quantum mechanics, was only a mathematical approximation for a relativistic quantum field theory. If the exact theory could not be demonstrated, it then had to be admitted that there was no guarantee that the nonrelativistic approximation came from a general theory based on any of the ideas of the Copenhagen school.

The reason for the fact that there are no solutions for the equations of relativistic quantum field theory is that infinite quantities automatically appear in its formulation. The predictions then follow that all of the physical properties of an elementary particle (its mass, electric charge, and so forth) have infinite magnitude. About 20 years after this discovery of a failure of quantum mechanics, a mathematical method was devised, called 'renormalization', whereby infinite quantities are subtracted from the original infinite quantities that appear in the formalism. In this way, numbers follow that are supposed to be the finite predicted properties of the elementary particles. In the case of quantum elecrodynamics (the form of relativistic quantum field theory that applies to the electromagnetic force in particular) this subtraction procedure yielded two remarkably close predictions (not even predicted qualitatively by ordinary nonrelativistic quantum mechanics, or a relativistic reformulation by Dirac). One of these was 'the Lamb shift' (a prediction of extra energy levels in the spectrum of hydrogen) and the other was a small added part to the magnetic moment of the electron (referred to as the 'anomalous magnetic moment of the electron'), anomalous compared with the prediction of Dirac's relativistic formulation of quantum mechanics (a formulation that was still incomplete because it had not yet added in the radiation field).

Even though the predictions of the Lamb shift and the anomalous magnetic moment of the electron were in very close agreement with the data (the measurement of the anomalous magnetic moment of the electron is the most accurately measured number in physics, to this date), it is still unfortunately the case that these results follow from a prescription that has never been shown to have mathematical consistency. This is an important point to make in regard to the authenticity of quantum field theory, proposed as a bona fide scientific theory. This is because of the rule of the philosophy of science, that it is a necessary condition that a proposed theory match the empirical data, but this is *not sufficient*. For a theory of science to be true it must also be logically consistent and it must predict unique answers for questions about unique physical situations. In its present state, quantum electrodynamics (and generally, quantum field theory) has not satisfied this criterion. Because of the nondemonstrability of the mathematical consistency in these schemes of renormalization, which utilizes perturbation methods whereby one expresses all physical observables in terms of power series expansions, *that diverge,* and then one subtracts off the infinite remainders of the series in some prescribed way, it is always possible to redefine the subtraction procedure by regrouping the series in a different way, thereby yielding different predictions for a given property.

The present-day widespread feeling among physicists that quantum mechanics has been overwhelmingly successful is based on the large amount of empirical confirmations it has given for the low energy data in the microdomain, as well as the extremely accurate predictions that have come from the renormalization procedure for some high energy results. However, it must still be admitted that there is yet no bona fide theory with demonstrable mathematical consistency. One must then not rule out the possibility that the general theory we seek, whose low energy approximation corresponds with nonrelativistic quantum mechanics, is a general theory that is not based on any of the concepts of the quantum theory, that is, the ideas listed in the left-hand column of Table 1. The possibility may then be considered that the general theory of matter that leads to quantum mechanics, as a particular approximation, may be based on the full conceptual content of the theory of relativity, the ideas shown on the right-hand side of Table 1. In this case, the imposition of *the principle of correspondence* would require that the nonlinear mathematical equations for the matter fields, according to this general approach, would incorporate quantum mechanics as *a linear approximation*. It will be proposed below that the *meaning* of the general matter equations will be to provide a fundamental explanation for the inertia of matter.

A further difficulty that implies incompatibility between the quantum and relativity theories is that the quantum theory *necessarily* entails an absolute frame of reference—the frame of the measuring apparatus. The eigenvalue form of quantum mechanics (the 'Hamiltonian form') implies an absolute simultaneity of both the act of measurement of some physical quantity (cause) and the revealed measured quantity (effect). This is then incompatible with relative simultaneity, because of the predicted finite speed of propagation of interaction, linking the cause and the effect *in a nonzero time* (in any, arbitrarily chosen, inertial frame of reference). That is, with the Hamiltonian formulation of quantum mechanics, if the cause and effect (the act of measurement and the revealing of this measurement) are expressed simultaneously in one Lorentz frame, relativity theory predicts that they will not be represented as simultaneous events in other Lorentz frames. But this would destroy the Hamiltonian eigenvalue form of the quantum mechanical equations in the other Lorentz frames. We see, then, that because of the role of measurement in quantum mechanics and its expression in Hamiltonian form, this theory is *manifestly* incompatible with the symmetry requirement of the theory of relativity, whenever the interaction of the apparatus and the measured micromatter is fully expressed, including the radiation that is transferred between them, when a 'quantum jump' occurs.

To sum up, it appears that the concept of the 'quantum jump' and its relations to a photon theory of light is i) logically incomplete, and ii) logically inconsistent. The answer to the question in the heading of this section is then: no, the 'quantum jump' is not compatible with the mathematical and conceptual bases of the theory of relativity.

Is the Theory of Relativity Complete as a Theory of Matter?

When examining the full conceptual basis of the theory of relativity as a fundamental theory of matter (with the ideas of the right-hand column of Table 1), it is seen that there is still something missing from the expression of this theory according to Einstein's formulation. This has to do with an explicit representation of the inertial manifestation of matter.

According to the meaning of *the principle of relativity,* one compares the expressions of the laws of nature in different reference frames, demanding that they should all be in one-to-one correspondence. But in order to make such comparisons, abstract spatial and temporal measures must be evoked, to be correlated with real readings of measuring instruments. These are instruments of any sort, whether they are the sophisticated instrumentation of modern day physics or the ordinary 'rods' and 'clocks' that were originally referred to in the early writings of Einstein and others on this subject. The point I wish to make here was originally made by Einstein himself, after a short period of reflection following his discoveries of special relativity: it is that, after all, these instruments are material entities and not theoretically self-sufficient expressions of space and time, in themselves.[89]

To express a basic theory of matter that would be self-consistent in accordance with the underpinnings of the theory of relativity there must be symmetry in the description between the field variables associated with the 'observer' and those of the 'observed'. It is then necessary *in principle* to express the matter field variables that relate to the physical behavior of the measuring instruments in a way that is compatible with the rules of the field concept in relativity theory, rather than merely referring to the instruments as 'rods' and 'clocks' that look down on matter as though they were totally disconnected 'outside observers'. Indeed, this is the view that the Copenhagen school takes about 'measurement': that it is disconnected, dynamically, from the measured. This is a view that is logically excluded by the basis of the theory of relativity.

Thus we see that the proper generalization of the theory of general relativity, to complete it as a fundamental theory of matter, would be to introduce the basic variables associated with the measuring instruments, as continuous (and analytic) fields that solve objective laws of nature: field equations that relate to the mutual interactions within a physically closed system of matter. Since the measurements must always entail the inertia of the matter of the measured (the 'emitter') as well as the inertia of the 'measurer', in principle (the 'absorber'), and since the symmetry of relativity theory requires that the interchange of the variables of the 'measurer' and the 'measured' must leave the complete description of this interaction unchanged in form, it follows that these field variables that describe the measurement process relate fundamentally to an explanation of the inertia of matter, and that this explanation is in terms of matter fields, rather than disconnected 'rods' and 'clocks'. Such a view, which is dictated by Einstein's *principle of relativity,* adheres to Bohr's insistence that the 'observer' must be incorporated with the 'observed' in any fundamental representation of matter. However, in contrast with Bohr's philosophy of asymmetry between the macroapparatus and the micromatter that is observed, the theory of relativity must introduce the 'observer' and the 'observed' symmetrically, in accordance with the spirit of Newton's third law of motion. Clearly, with this approach there would be no difference between macrovariables and microvariables—they must both be variables of the same sort, that is, 'covariant' fields in general relativity (fields that have the same form in any frame of reference).

We see, then, that the conceptual requirement of relativity theory, that there be symmetry between the variables of the 'observer' and the 'observed', logically requires a continuous field theory, because of the distinguishability of reference frames in terms of continuous transformations of the space and time coordinates, and that the continuous matter fields must relate to a fundamental *explanation* of the inertial manifestation of matter, rather than simply inserting a mass parameter to fit the data. This continuous field theory of inertia must then unify with the force manifestations of matter (gravity, electromagnetic interactions, nuclear and weak interactions, and so forth) in a single self-consistent set of field equations in general relativity. The demonstration of such a theory is given in my *General Relativity and Matter* (see note 74). This is the *unified field theory* that Einstein sought for the greatest part of his professional career. His motivation was not purely esthetic; neither was it due to his genuine desire for maximum simplicity, or to some not-too-well-understood intuitive reasons. It was based primarily on his recognition of the *logical necessity* for such a unified field theory, if the theory of relativity is to claim to be valid as the basis for a general theory of matter.

Concluding Remarks

Summing up, it has been argued in this chapter that neither the quantum theory nor the theory of relativity are in themselves complete as fundamental theories of matter. Quantum mechanics is not complete because its nonrelativistic approximation has never been satisfactorily generalized relativistically, so as to accommodate the 'quantum jump' (one of the essential features of this theory of matter) in the form of a relativistic quantum field theory. The theory of relativity, on the other hand, is not complete in Einstein's original formulation because it does not yet incorporate the field variables of the 'observer' in a fully 'covariant' manner, in terms of a field description of the 'observer-observed' interaction, including the inertial manifestation of matter. It was seen that the inertial manifestation must be unified with the rest of the field theory in order to provide a complete description of the measurement process—*one of the interactions* that must be incorporated with all of the other possible interactions of a closed system. This is in contrast with Bohr's philosophy, which asserts that the measurement interaction is all that there is to talk about.

I have discussed the reasons for the failure of the quantum theory from the theoretical view (in spite of its empirical successes) to achieve a complete description. The completion of the theory of relativity does not suffer from the same conceptual or mathematical difficulties. But it is clear that to proceed from this viewpoint it is necessary to totally abandon the conceptual basis of the Copenhagen view of quantum mechanics, while keeping its form in terms of a probability calculus as a nonrelativistic approximation for an exact field theory of inertia, that is not at all, in its general form, a probability theory.

If it turns out, in the long run, that the theory of general relativity will indeed replace the quantum theory as a fundamental theory of matter, then one might view the historical evolution of theories of matter as a progression from the idea of 'particle monism', of the periods preceding the twentieth century, particularly the ideas of Newtonian physics, and earlier, the Greek ideas of atomism, to the period of the first half of the twentieth century, where the concept of 'wave-particle dualism' of the quantum theory was held to be true, thence to a view of (nonlinear) 'wave monism', rooted in the theory of general relativity, as a fundamental theory of matter.

This progression of the history of physics would then imply that the dualistic concept of wave and particle was not permanent in our accumulated understanding, but that it did serve the very useful purpose of the intermediate role of superposing the earlier (concrete) concept of atomism with the (more abstract) holistic concept that came with the continuous field view as fundamental, and in paving the way for the emergence of the latter fundamental approach to matter.

Such a replacement, while new in physics, is certainly not new in other areas of the history of ideas. The holistic idea of the material world may indeed be traced back to ancient times in both the Oriental and the Western cultures. In the Orient, the holistic view was taken as long ago as about 3000 years in the ancient Brahmanical and Buddhist views in India, and in ancient China, in the writings of the Taoists, such as Lao Tzu and his disciples and the various offshoots of this philosophical approach. In Western culture, starting also about 3000 years ago in the philosophical writings and ideas carried by word of mouth, there is the holistic approach in ancient Greece of Heraclitus, Parmenides and Plato, the writings from Judaic culture of the Kabbalists, and later on, in the seventeenth century writings of Spinoza. It is my opinion that indeed Spinoza's philosophy is the most important philosophical precursor for the approach of Einstein's theory of general relativity.

My own research program over the past 30 years has taken the holistic, continuum approach, based on the theory of general relativity. From the results of these theoretical studies, I believe that one may satisfy Einstein's criteria when sufficient generalizations have been implemented. One of these extensions, as we have discussed in the preceding chapter, is the generalization of the *Mach principle* to *all* of the manifestations of matter, in addition to its inertial manifestations. All of these physical manifestations of elementary matter are then to be taken as not more than a degree of coupling within a physically closed continuum. In this way, the atomistic model is totally disposed of, and the inertial manifestations of elementary matter are seen to fully fuse with its force manifestations in the form of a unified field theory, based entirely on the continuous field concept. It was found in this research program that the part of the field theory that explicitly provides the explanation for the inertial manifestation of elementary matter—a set of nonlinear (spinor) field equations in general relativity—approaches the form of nonrelativistic (linear) quantum mechanics, in the appropriate limit of sufficiently small energy-momentum transfer between the interacting components of a material system. Thus it was found that all of the successful results of quantum mechanics, from the point of view of predicting the empirical evidence, as well as the quantum electrodynamical prediction of the Lamb shift, are contained in the predictions of this theory, though without any dependence on any of the assumptions of the quantum theory, according to the Copenhagen interpretation.[90] In addition, many more predictions have been made that have never been made in a mathematically satisfactory way (or at all) from the basis of the quantum theory. Thus, it is my belief that, indeed, it was Einstein who was right after all, in his debates with Bohr as to whether the basis of the relativity approach or that of the quantum approach is more true to nature.

On the other hand, scientific truth is only provisional; thus it might very well turn out in the final analysis that more investigation will reveal that neither Bohr nor Einstein was fully right and that there is a new view in physics that will largely replace both of theirs. If this should be the case, I still feel that most likely the *principle of relativity* of Einstein's theory is a concept that will remain in any future development of physics, even if other features of his theory may wither away. I personally believe this simply because I find it very hard to imagine that this principle of nature could be false, if only because of its extreme simplicity. At the same time, I fully admit that this is a personal intuitive feeling of one scientist, and that other scientific views could in principle replace this one in some future scientific era.

Postscript

The primary aim of this monograph has been to expose the conflicts between the underlying ideas of the two primary revolutions in twentieth-century physics—the quantum and relativity theories. The dilemma that has been reached in the present stage of physics is that, on the one hand, each of these theories needs the other for its completion *on its own terms,* but on the other hand, since there are logical dichotomies in the union of the axioms of each of these theories, they cannot be logically unified. The implication of this state of affairs is that we will achieve real progress in our understanding of the nature of elementary matter only if we either abandon the basis of one of these theories for the other, or else abandon both theories for an entirely new (yet to be discovered) view. Taking the former stance, it must still be necessary to recover the successful mathematical results of the abandoned theory. The proposal in this monograph is that a reasonable way to do this would be to require that a proper generalization of the maintained theory should have a formal expression that may be mathematically approximated by the formal expression of the abandoned theory, in the appropriate limits where the latter had been empirically successful.

Since the initial successes of the quantum theory, the great majority of physicists have been convinced that (with perhaps minor modifications) the view of the Copenhagen school is true to nature, and will certainly withstand the test of time. Perhaps this is a correct judgement. Yet it has not yet proven itself to be so, conclusively, for several technical reasons, as we have discussed in this book. Thus, it may still be admitted that there is a finite possibility that this is a false judgement, even though some very brilliant scientists believe it to this day.

If, perchance, the quantum theory is not a true underlying explanation of the nature of micromatter, a possible alternative is a theory based on the ideas of general relativity theory, as I have discussed in this book. What I have argued for is that if this theory is indeed closer to a valid approach to the nature of matter, then it must still incorporate the formal structure of the quantum theory in a particular limit. This is an assumption of the well-known *principle of correspondence.* This principle, which has been followed throughout the history of science since ancient times, states that when a new theory supersedes an older one, its mathematical form must go smoothly into the earlier form that had been empirically successful, as a particular approximation that is appropriate under the conditions where the older theory had been mathematically successful. An important example to demonstrate this is the case of Einstein's theory of general relativity superseding Newton's theory of universal gravitation. Each of these theories of the gravitational force is *entirely* different, both conceptually and mathematically. Still, when distances and relative speeds of interacting masses are sufficiently small, Einstein's field theory takes the form of Newton's theory, as a close mathematical approximation.

Thus, if it is true that in the debates between Bohr and Einstein, that it was Einstein who was right after all, then we would have to say that the basic approach to a fundamental explanation of matter (in any domain from the microscopic to the cosmological) is rooted in *Einstein's principle of general relativity.* If we reflect further that the *generalized Mach principle* must be added to the axiomatic basis of a general theory of matter in general relativity, as we have discussed in chapter ten, as well as the *principle of correspondence,* then these three basic axioms yield the structure of a general theory of matter that, in a certain approximation, leads to the formal expression of quantum mechanics. But the *meaning* of the quantum mechanical equations in this approximation is not at all the same as that proposed by the Copenhagen school. Rather, quantum mechanics becomes a nonrelativistic, linear approximation for a field theory of inertia, expressed in general relativity.

In this way, we see that the ideas of indeterminism, fundamental probability and linearity that are logically required in the quantum 'particle' theory are replaced by the concepts of determinism and total objectivity in a theory of matter based on the continuous field concept.

It is my opinion that the evolution of physics in the near future will be based on the foundations of general relativity, as a fundamental theory of matter, rather than the present-day views of the Copenhagen School. Thus, I do believe that it was Einstein who was right, as were a few others in their disagreement with the Bohr school, such as Schrödinger and Planck. But, of course, we are only finite creatures so that we can never know with certainty how our understanding will evolve in the future, no matter how sure we may feel about the current explanations. All that we can do, as individual scientists, is to follow our own intuitive leanings, whether or not they agree with the majority of the authorities at any given time. It is my hope, in presenting the discussions of this book on the controversies in modern physics, along with discussions of what these theories are really all about, from a conceptual point of view, that the reader will get the flavor of scientific thinking in its struggle for genuine progress in our understanding of the material world. I hope also that the reader will feel free to use his or her own intuitive feelings as to which set of concepts is more likely to be closer to the real world. And I hope that the reader will gain the sense of excitement and adventure that is always present in the search for scientific knowledge.

Implications of Holism: From Tao to Relativity Physics

I would now like to speculate on a generalization of Einstein's principle of relativity, which in its initial form has been so successful in leading to an understanding of inanimate matter. For if it is, indeed, Einstein's approach that will win out over its twentieth-century competitor, the quantum theory, as a fundamental theory of matter, then the generalization proposed may be significant for the later development of ideas about the real world, perhaps not to come to fruition until the next century.

As we have seen, the theory of general relativity is based, axiomatically, on the principle of relativity, which, in turn, may be said to express the notion that the truths of the world (the laws of nature) applied to inanimate matter, must be totally objective. That is, the laws of inanimate matter are said to be in *one-to-one correspondence* in all possible frames of reference.

A natural generalization of this principle, which has already had great success in our understanding of the material world, from the domain of elementary particle physics to that of cosmology, is that it should also apply to the other laws of nature, that transcend the domain of knowledge normally referred to as 'physics'. That is, I propose, as a generalization of Einstein's principle of relativity, that the truths in *any* domain of knowledge should be invariant. Such extended domains of knowledge are the philosophical categories (metaphysics, epistemology, ethics, aesthetics, and so forth), the categories of the social sciences (psychology, archaeology, anthropology, sociology, and so forth), the life sciences (biology, physiology, and so forth). If some assertion of a (contingent) truth about the world is indeed a valid description of an objective feature of the world, then it should be found in all of the categories of knowledge, though expressed in different contexts, in different languages, and within different cultural reference-frames, at different times and places.

If the truths of the world are indeed invariant to transformations from one intellectual discipline to any other, in this way, from one cultural context to any other, then it behooves those who seek truth to study the abstract features of the truths of as many disciplines as possible, in order to determine which of the ideas of each of them correspond and which do not, with the notion that those ideas that do recur in a varied range of domains of knowledge are more likely to be true than those that do not. Thus the seemingly invariant truths are the ones that should be pursued further, as significant investigations toward our future understanding of the real world.

If this statement about the invariance of the truths of the

world is itself true, then it must have been expressed many times in the past, in different ways, in different cultures and in different contexts and languages, as it will continue to be repeated in the future. This is indeed the case. For example, in the ancient writings of the Hindu culture of three thousand years ago, it is said that[91]

> Truth is never lost, but may be undiscovered or forgotten. It can therefore be discovered and recovered.

The implication here is that there are indeed invariant truths of the world, and when it appears that a *new* truth has been discovered, in any domain of knowledge, this is in reality a *timeless* truth that has, rather, been recovered (perhaps in new 'clothing').

Another example of expression of this particular idea comes from the Biblical scriptures, where it is said in the Book of *Kohelet (Ecclesiastes):*

> The thing that hath been is that which shall be; and that which is done is that which shall be done; and there is no new thing under the sun.

With the thesis of this generalization of Einstein's principle of relativity to all other possible domains of knowledge, I will now briefly discuss the application of this idea to another particular truth (that of holism) and I will attempt to show that this single concept has been indicated in different cultures and at different times and places, expressed in different contexts, throughout the ages. This implies that, in contrast with the various atomistic views, it is perhaps the holistic view that is an invariant truth of the real world.

Two holistic views from diverse cultures that I will now compare are in the contexts of 1. Hua-yen Buddhism[92]—a Chinese version of the Buddhistic teachings that had moved from India to China, and somewhat influenced by the ideas of Taoism of Lao Tzu,[93] and 2. the contemporary *I-Thou* philosophy of Martin Buber[94]—a view that has some of its roots in ideas of Jewish mysticism (the *Kabbalah*) which, of course, also emerged from ancient times.

A fundamental concept of the original Indian Buddhism that

carried over to the Buddhistic view adopted in China is that of *anātman* (translated from the Sanskrit as 'nonbeing'). The opposite concept, the idea that the essence of matter is in being (ātman) would signify that the objective world is indeed made out of separate things. In Buddhism, instead, the objective material world is characterized most fundamentally by 'becoming'. (It is interesting to note, in the context of comparative studies, that in the Kabbalah of Jewish philosophy, one views the human being as in continual becoming, recreated at each instant as the world evolves.)

This view of 'becoming', as essence, is comparable with the field concept in Einstein's general relativity theory, as we have discussed earlier in the text. For, in this theory of matter, the field solutions of the laws of nature do not represent a 'thing'; rather, they have *dynamical* significance. A dynamical system expresses the cause-effect relations in space and time. Thus the elementarity of 'becoming' in Buddhism is conceptually closer to the interpretation of Einstein's matter field, as the essence of matter, than is 'being'.[95] But in relativity theory this dynamics does not entail the history of a single entity. It rather relates to a single closed system with an indefinite number of related components that are *not separable*. That is, no component of this system is meaningful as a thing by itself; its only meaning is in terms of the *totality* of the closed system. For the field theory that follows from Einstein's relativity concept leads to the elementarity of relation rather than relata.

The view of totality as foremost in the final analysis is very similar to the basic ontology of Hua-yen Buddhism.[96] This view has been discussed at length by scholars of Chinese Buddhism, such as Cook. The concept of oneness then enters in relativity theory of the contemporary period in theoretical physics and in the cosmology of the ancient writings on Hua-yen Buddhism, in the form of *totality* of a closed system of (irremovable) components that comprise the material universe. Illustrating this view, Cook said, in the closing sentence of his book: "It is not just that we are all in it, together. We all *are* it, rising or falling as one living body."

This idea of oneness of the universe is similar to a particular interpretation of Martin Buber's *I-Thou* philosophy.[97] In

Buber's own view, it is the existential relation, *I-Thou,* which is fundamental and irreversible rather than the separate relata. From my reading of Buber, the elementarity of *I-Thou* eliminates any semblance of an absolute subject, *I,* relating to an absolute object, *Thou.* Rather, all that exists is a single object, *I-Thou* wherein the hyphen is in principle irremovable. In my view, the *I-Thou* relation is a single, inseparable whole, perhaps similar to the *tao* of Lao Tzu, in ancient Chinese philosophy.[98]

In contrast with my reading of Buber, his own interpretation of *I-Thou* is in terms of two entities rather than one. From his view, the existence of the consciousness allows that the *I* (subject) may totally extricate itself from the rest of all that there is (the rest of the universe and God), leading, residually, to the rest of the world as an object, set in the context of space, time and matter—the *it.* Only then does *I* become aware of *it* as 'other', an object to comprehend. Indeed, the existence of the *I-it* relation is necessary for us to acquire any scientific understanding of the world.

My dialogue with Buber does not concern whether or not an *I-it* relation truly exists. In my view, it certainly does exist, in the same sense that the Cartesian thinker exists *(cogito ergo sum).* My question is, rather, *how* does the *I-it* relation emerge from the *I-Thou* relation? I see two alternative answers to this question. The first is Buber's own intention, that *I* and *Thou* are logically distinct entities, wherein *I,* at will, enters into a relation with *Thou,* and simultaneously, *Thou* enters into a relation with *I,* as separate Person and person, though in mutual relation. This view is clearly expressed in Buber's writing.[99]

> The description of God as a Person is indispensable for everyone who means by 'God', Him who—whatever else He may be—enters into a direct relation with us men in direct creative revealing and redeeming acts, and thus makes it possible for us to enter into a direct relation with Him.

Buber thus insists, with this view of *I-Thou,* that along with the usual attributes of God, one must add that of 'Person', an individuated self, insofar as He participates in an *I-Thou* relation with individual consciousnesses.

An alternative interpretation of the *I-Thou* relation is that it is truly *one,* the whole that continuously is one and nonseparable, the existent that is all that there is. With this interpretation, then, *Thou* refers to an aspect of a *closed relation, I-Thou:* a relation that entails a transcendent God, and the natural universe that is, in turn, one of His manifestations. One of the infinitude of His manifestations, that comprises the natural universe, is the subject *I.* That is to say, in this case *I* is a relative (rather than an absolute) *subject* of the *I-Thou* relation, focussing on the individual consciousness and its ability to be aware of 'other'.

The question then arises: *how* does an individual consciousness come into play if *I-Thou* is indeed one? An answer is that, in the holistic interpretation, *I* appears in the sense of an *approximation* that an aspect of the whole, called 'human consciousness', is capable of establishing, making it *appear* that there are truly two things—an absolute subject, *I,* and an absolute object, *it.* Still, this is an approximation that only gives the illusion of two, whereas *I-Thou* is really one, a single entity. That is to say, with Buber's philosophy, using this alternative interpretation of his words, the *I-it* relation is not on the same ontological basis as *I-Thou*; rather, the *apparent* objectivity of *I-it* refers to *an idea of* a component manifestation of the whole, the *I,* reached from 'individual' reflection. It is only in the latter state of mind that *I* can have awareness of 'other', as is required to understand the world in terms of a scientific representation. Thus, it only *seems* to the 'observer', *I,* when in this state (of the whole), that it is in itself an absolute subject of the *I-it* relation—an 'observer' of the absolute object, *it.*

Thus we see that, from Buber's own standpoint, there are two relations that are complementary: *I-Thou* and *I-it.* When one is, the other cannot be. Yet, the subject, *I,* may choose at will to enter into one of these relations or the other. This type of complementarity is, perhaps, along the subjectivist philosophical line, similar to Bohr's principle of complementarity in the modern day quantum theory.[100] While it is still *relation* that is elementary in Buber's philosophy, and while *I-Thou* is a single objective relation and *I-it* is a subject-object relation, *I-Thou* and *I-it* are on ontologically equivalent levels, according to Buber himself.

The other alternative to an interpretation of Buber's philosophy asserts that *I-Thou* is all that there really is. It is continuously whole and nonseparable, possibly similar, conceptually, to the *tao* of Lao Tzu of ancient Chinese philosophy,[101] or the Brahman of Hinduism.[102] Here, all that one may say about the *I-Thou* is that *it is*. It is not defined within the context of space, time or matter. It is ontologically prior to all other things. *As an approximation,* the human consciousness, as a mode of the whole, may distance itself from the remainder, thereby becoming aware of *it,* of the *I-it* relation. Still, *I* is not to be regarded, with this view, as an absolute subject, separable from the *I-Thou.* Rather, it is a manifestation of the whole—the *I-Thou*—seemingly giving the consciousness an impression of a world 'out there'.

In modern physics, the philosophical approach of Einstein's theory of general relativity, as an underlying theory of matter, is closer to the latter view of *I-Thou,* as a truly closed system, while the former interpretation of *I-Thou/I-it,* as complementary relations, is closer to Bohr's philosophy of complementarity, according to the present day quantum theory. It is interesting, then, that the conflict between these possible alternative interpretations of Martin Buber's *I-Thou* philosophy seems to compare with contemporary conflict between Bohr's and Einstein's approaches to the truth of the physical universe.[103] The latter conflict of modern times may be compared, in turn, with the debates on holism versus particularity which have occurred since ancient times.

According to Einstein's interpretation of his theory of relativity, the universe must be characterized as a closed system, continuously distributed everywhere. Thus, there can be no *localized* absolute subject, *I,* looking down on the universe, so to speak, as another *localized* object. With this view, the human consciousness itself is not more than one of an infinitude of manifestations of the single universe, that is continuously one and nonseparable. The role of the single manifestation *of* the universe (rather than a thing in it) is analogous to an (internally created) ripple *of* a pond. Indeed, the ripple is not a separable thing *in* the pond, that could be removed and studied on its own, say by measuring its weight, size, color, and so forth. Rather, the ripple is not more than a mode of behavior of the entire pond: it is *of* the pond. As we come to understand the ripple more fully, we see that it reflects the nature of *the entire pond, holistically*. In the same way, the human being may reflect on the entire universe, understanding his own role as not more than one of its infinite number of manifestations of the continuous closed system that is the universe, according to Einstein's general relativity field ontology in modern physics, or the holistic views of the Asian scholars and the Greek scholars of the ancient times.

The latter cosmology seems to me similar also to that of Brahmanism in ancient Hindu culture. In this view, the single reality, all that there is, is *Brahman*. Here, the essence of being, the *ātman*, is one with *Brahman*. While this ancient view is much older than the explicit expression of the *I-Thou* existentialist view, Buber's ideas were strongly influenced by ancient ideas of Jewish mysticism—the *Kabbalah*. This was not expressed in written form (in the *Zohar*) until the Middle Ages; it is said, however, to have been expressed orally since biblical days, passed on from one generation to the next. Such an argument is given by Franck.[104]

The Brahmanical view, similar to the view of Spinoza, extends from a naive sort of pantheism to the theory that the natural universe itself is only one of an infinite manifold of the manifestations of God: that is, that God transcends the natural universe (the subject of physics). Though the natural universe, in this view, is not separate from God, it is not all that there is to God. One sees this view in the *Bhagavadgita,* of Hindu literature, where it is said of *Brahman:*[105] "Undivided, He seems to divide into objects and creatures. Sending creation forth from Himself."

One may find a very similar passage in the Kabbalah of Jewish philosophy, where it is said:[106]

> He is the beginning and the end of all degrees of

> Creation; all degrees are worked with His seal, and He can be designated only by unity. He is *one* despite the innumerable forms with which He is invested.

Again in Western culture, several centuries later one finds this cosmological view of holism in ancient Greece, in Plato and Parmenides, and then in the Hellenistic period, in Plotinus, who said:[107]

> All things that exist do so by virtue of "unity"—in so far as they exist in any ultimate sense and in so far as they may be said to be real. . . . But of this One no description nor scientific knowledge is possible. . . . And it is wholly self-sufficient by virtue of its being simple and prior to all things.

I have tried to demonstrate in the preceding brief comments that the cosmological approach of oneness that underlies the Asian cultures is also embedded in views of the Western culture, from Biblical times to ancient Greece to Spinoza and to the modern era of Buber and Einstein.

It is interesting to note that at the present stage of the history of science, a stalemate has been reached in regard to the two 'revolutions' of twentieth-century physics. The quantum theory and the theory of relativity take opposite sides on the question of particularity versus continuity and holism (where the term 'revolution' is only meant in the sense of a rapidly changing stage of evolution).

The quantum and relativity theories are fundamentally incompatible with each other at the conceptual level, as well as in regard to their respective mathematical expressions, as discussed in chapter 10. This is because of the contrasting views that one may associate with atomism, indeterminism, and positivism, versus continuity, determinism, and realism. The dilemma that appeared was rooted in the fact that the quantum theory, *on its own terms,* must necessarily fuse with the theory of relativity (in the form of a 'relativistic quantum field theory'). But such a fused theory has never been constructed in a logically and mathematically satisfactory manner, not since the onset of quantum mechanics in the 1920s. One may then conclude that the future of physics will lie in the direction of one of these theories (suitably generalized) while rejecting the other. My choice, for several technical as well as intuitive reasons, is that the basis of Einstein's general relativity will survive, as an elementary theory of matter, in all domains from 'particle physics' to cosmology. This turn away from the basis of the quantum theory must nevertheless keep the nonrelativistic approximation of quantum mechanics as a useful calculational device in the domain of the low energy physics of micromatter, thus adhering to the *principle of correspondence,* which has applied thoughout the history of science.

This conclusion would then bring us back in our understanding of matter to the monism of the continuous field concept in general relativity, a philosophical view of holism, and oneness.

Fully incorporating the holistic philosophy would imply a natural extension of the basis of Einstein's theory of general relativity so as to include humankind and consciousness, a theory perhaps similar in outlook to the ancient writings from India and China, more than 25 centuries ago, and to the ideas of Parmenides and Plato, the Jewish Kabbalists, and similar conceptions down through Spinoza, Buber, and Einstein. Because there is this overlap of ideas from such diverse cultures, at different times and places and expressed in different languages and in terms of different contexts, perhaps there is more chance that there is truth in this approach to the universe than in the opposite atomistic views that have dominated much of Western philosophy until now.

Notes

1. From 'A Talk with Einstein', *The Listener* (September 1955).
2. From Niels Bohr, 'Can Quantum Mechanical Description of Physical Reality be Considered Complete?', *Physical Review* 48 (1935), p. 696.
3. For opposing views on the subject of scientific revolutions, see: I. Lakatos and A. Musgrave, eds., *Criticism and the Growth of Knowledge* (Cambridge: Cambridge University Press, 1972); T.S. Kuhn, *The Structure of Scientific Revolutions* (Chicago: University of Chicago Press, 1970).
4. I have elaborated on this concept in: M. Sachs, *Quantum Mechanics from General Relativity* (Dordrecht: D. Reidel Publishing Company, 1986), chapter 2.
5. The full text of this prayer is given in: J.S. Minkin, *The World of Moses Maimonides* (New York: Thomas Yoseloff, 1957), p. 149.
6. English translations of Galileo's main works are by Stillman Drake: *Dialogue Concerning Two Chief World Systems,* second edition (Berkeley, California: University of California Press, 1970), and *Two New Sciences* (Madison, Wisconsin: University of Wisconsin Press, 1974).
7. An excellent discussion on the arithmetic relations of numbers is in C. Lanczos, *Numbers Without End* (Edinburgh: Oliver and Boyd, 1968).
8. E. Mach, *The Science of Mechanics* (La Salle, Illinois: Open Court, 1960), p. 168, and Drake, *Two Chief World Systems* (see n. 6 above).
9. N. Bohr emphasized the importance in science of the *principle of correspondence* in his discussions of the

relation of quantum mechanics to classical mechanics. See: M. Jammer, *The Conceptual Development of Quantum Mechanics* (New York: McGraw-Hill, 1966), sec. 3.2.
10. Newton's original discussion of motion is in: I. Newton, *Principia,* vol. I, trans. A. Motte, revised by F. Cajori (Berkeley, California: University of California Press, 1962).
11. Mach, *The Science of Mechanics* (see n. 8 above), p. 264.
12. The history of Kepler's researches is well described in: A. Koestler, *The Sleepwalkers* (Harmondsworth, U.K.: Penguin, 1964).
13. Newton's discussion of optical phenomena is in: I. Newton, *Optiks* fourth edition (New York: Dover Publications, 1952). For further discussion, see: M. Hesse, *Forces and Fields* (Edinburgh: Thomas Nelson & Sons, 1961).
14. An account of Huygens' work is in his book: *Treatise on Light* (London: Macmillan, 1912), trans. S.P. Thompson.
15. E.S. Haldane and G.T.R. Ross, trans., *The Philosophical Works of Descartes* (Cambridge: Cambridge University Press, 1934). Also see: B. Williams, 'Descartes', in *Encyclopedia of Philosophy,* vol. II (New York: Macmillan, 1967), p. 344.
16. R. Descartes, 'Meditations on First Philosophy', translation in: M.C. Beardsley, *The European Philosophers from Descartes to Nietzsche* (New York: The Modern Library, 1960).
17. B. Spinoza, *The Principles of Descartes' Philosophy* (La Salle, Illinois: Open Court, 1905).
18. The essence of Spinoza's philosophy is in: B. Spinoza, *Ethics,* trans. R.H.M. Elwes (New York: Dover, 1955).
19. For a clear discussion of the influence of Maimonides on Spinoza, see: H.A. Wolfson, *The Philosophy of Spinoza,* vol. II (New York: Schocken Books, 1969). Also see, M. Sachs, 'Maimonodes, Spinoza and the Field Concept in Physics', *Journal of the History of Ideas* 37 (1976), p. 125.
20. N.M. Glatzer, *The Judaic Tradition* (Boston: Beacon, 1969), p. 279.
21. *Ibid,* p. 419.
22. D.G.C. MacNabb, 'David Hume', in *Encyclopedia of Philosophy,* vol. III (New York: Macmillan, 1967), p. 74.
23. W.H. Walsh, 'Immanuel Kant', in *Encyclopedia of Philosophy,* vol. III (New York: Macmillan, 1967), p. 305.
24. For an interesting discussion of the history of conservation

laws in physics, see: Y. Elkana, 'The Conservation of Energy: A Case of Simultaneous Discovery?', *Archives d'Histoire des Sciences,* no. 90 (1970), p. 31.
25. For a discussion of the laws of thermodynamics, see: R.P. Feynman, R.B. Leighton and M. Sands, *Lectures in Physics,* vol. 1 (Reading, Massachusetts: Addison-Wesley, 1963), chapter 44.
26. P.K. Feyerabend, 'Ludwig Boltzmann', in *Encyclopedia of Philosophy,* vol. I (New York: Macmillan, 1967), p. 334. Also see: Y. Elkana, 'Boltzmann's Scientific Research Programme and its Alternatives', *Proc. Symposium Van Leer Jerusalem Foundation,* (January, 1971). Boltzmann's approach to theoretical physics was presented in a lecture in 1899, published in translated form recently as: L. Boltzmann, 'On the Development of the Methods of Theoretical Physics in Recent Times', trans. Y. Elkana, *The Philosophical Forum* 1 (Fall, 1968).
27. A standard treatise on the subject of statistical mechanics that treats these applications rigorously is: R.C. Tolman, *The Principles of Statistical Mechanics* (Oxford: Oxford University Press, 1938).
28. B. Russell, *An Inquiry into Meaning and Truth* (1940), quoted in A. Einstein, 'Remarks on Bertrand Russell's Theory of Knowledge', in P.A. Schilpp, ed., *The Philosophy of Bertrand Russell,* (La Salle, Illinois: Open Court, 1944), p. 283.
29. K.R. Popper, *Conjectures and Refutations* (New York: Harper Torchbooks, 1963).
30. Mach, *The Science of Mechanics,* (see n. 8 above), p. 342.
31. M. Sachs, 'Positivism, Realism and Existentialism in Mach's Influence on Contemporary Physics', *Philosophy and Phenomenological Research* 30 (1970), p. 403.
32. A.J. Ayer, *Language, Truth and Logic* (New York: Dover, 1952).
33. J. Agassi, *Faraday as a Natural Philosopher* (Chicago: University of Chicago Press, 1971). Faraday's original researches were reported in: M. Faraday, *Experimental Researches in Chemistry and Physics,* vols. I., II, and III (London: Taylor and Francis, 1859).
34. M. Sachs, *The Field Concept in Contemporary Science* (Springfield: Charles C. Thomas, 1973).
35. Maxwell's major discoveries in the subject of

electromagnetism are in: J.C. Maxwell, *A Treatise on Electricity and Magnetism,* vols. I and II (New York: Dover, 1954). A critical examination of Maxwell's theory is in: A. O'Rahilly, *Electromagnetic Theory,* vols. I and II (New York: Dover, 1965). Also see: R.A.R. Tricker, *The Contributions of Faraday and Maxwell to Electrical Science* (Oxford: Pergamon, 1966).

36. A. Fresnel, *Annales de Chimie et de Physique* 11 (1819) pp. 246, 377.
37. R.B. Leighton, *Principles of Modern Physics* (New York: McGraw-Hill, 1959), p. 4.
38. J.J. Thomson, 'Cathode Rays', *Philosophical Magazine* 44 (1897) p. 293.
39. M. Planck, *Theory of Heat Radiation* (New York: Dover, 1959).
40. Einstein's notion of wave-particle dualism first arose in his analysis of the photoelectric effect. Its history is discussed in: Jammer, *The Conceptual Development* (see n. 9 above), sec. 1.3.
41. It was G.N. Lewis who named the quanta of electromagnetic radiation, 'photons' (much later than their discovery). See: G.N. Lewis, 'The Conservation of Photons', *Nature* 118 (1926), p. 874.
42. This problem was addressed in the paper by N. Bohr, H.A. Kramers and J.C. Slater, 'The Quantum Theory of Radiation', *Philosophical Magazine* 47 (1924), p. 785. The attempt was made to consider the conservation laws to be valid only in a statistical sense, thus valid on the average over long periods, but not at any specific time.
43. W. Ritz, 'On a New Law of Series Spectra', *Astrophysical Journal* 28 (1908), p. 237.
44. A summary of this work is in: L. de Broglie, *Recherches d'Un Demi-Siècle* (Paris: Albin Michel, 1976).
45. C.J. Davisson and L.H. Germer, 'Diffraction of Electrons by a Crystal of Nickel', *Physical Review* 30 (1927), p. 705.
46. G.P. Thomson, 'Experiments on the Diffraction of Cathode Rays', *Prod. Roy. Soc.* (London), A117 (1928), p. 600.
47. The details of Schrödinger's derivation of wave mechanics are shown in: M. Sachs, *Ideas of Matter* (Washington: University Press of America, 1981), pp. 101–107.
48. For further discussion, see: M. Jammer, *The Philosophy of Quantum Mechanics* (New York: John Wiley and Sons, 1974), sec. 2.2.

49. *Ibid.*, sec. 2.3.
50. Some of the pioneering work of Max Born is presented and discussed in: B.L. van der Waerden, ed., *Sources of Quantum Mechanics* (New York: Dover, 1968). Born's views are also clearly expressed in: P.A. Schilpp, ed., *Albert Einstein: Philosopher-Scientist* (La Salle, Illinois: Open Court, 1949), p. 163.
51. L. de Broglie, *The Current Interpretation of Wave Mechanics* (New York: Elsevier Publishing, 1964), chapters 4, 6.
52. K.R. Popper, *Quantum Theory and the Schism in Physics* (Totowa, New Jersey: Rowan and Littlefield, 1982), secs. 22–28.
53. Boltzmann, 'On the Development of Methods' (see n. 26 above).
54. W. Heisenberg, 'Über quanten theoretische Umbedeutung kinematischen und mechanischen Beziehungen', *Zeitschrift für Physik* 33 (1925) 879. An English translation, 'Quantum Theoretical Re-Interpretation of Kinematic and Mechanical Relations', is in van der Waerden, ed., *Sources of Quantum Mechanics* (see n. 50 above), p. 261.
55. An accounting of Schrödinger's attempts to formulate a relativistic form of wave mechanics (which is the expression he sought *before* going to nonrelativistic wave mechanics) is given by: P.A.M. Dirac, 'The Evolution of the Physicists' Picture of Nature', *Scientific American* 208 (1963), p. 45.
56. M. Born and P. Jordan, 'Zur Quantenmeckanik', *Zeitschrift für Physik* 34 (1925), p. 858. An English translation, 'On Quantum Mechanics', is in van der Waerden, ed., *Sources of Quantum Mechanics* (see n. 50 above), p. 277.
57. P.A.M. Dirac, 'Foundations of Quantum Mechanics', *Nature* 203 (1964), p. 115.
58. A proof is demonstrated in: M. Born, *Atomic Physics,* fourth edition (New York: Hafner Publishing Company, 1946), appendix 22.
59. This example of a thought experiment for the demonstration of Heisenberg's uncertainty principle is similar to the one given in: G. Gamow, *Thirty Years that Shook Physics* (Garden City, New York: Doubleday, 1966), p. 107.
60. Bohr's reply to the Einstein 'photon box experiment' is in: N. Bohr, 'Discussion with Einstein on Epistemological

Problems in Atomic Physics', in Schilpp, ed., *Albert Einstein: Philosopher-Scientist* (see n. 50 above), p. 226.

61. R.P. Feynman, R.B. Leighton and M. Sands, *The Feynman Lectures in Physics* (Reading, Massachusetts: Addison-Wesley, 1963), chapter 37.
62. For a philosophical discussion of the concept of mass, see: H. Weyl, *Philosophy of Natural Science and Mathematics* (Princeton: Princeton University Press, 1949), chapter 22. The view of matter in ancient Chinese philosophy is in: W.-T. Chan, *A Source Book in Chinese Philosophy* (Princeton: Princeton University Press, 1963), pp. 339, 353, 420.
63. A. Einstein, B. Podolsky and N. Rosen, 'Can Quantum Mechanical Description of Physical Reality be Considered Complete?', *Physical Review* 47 (1935), p. 777.

    It has recently been revealed from his personal correspondences (with Erwin Schrödinger) that Einstein did not directly collaborate in the writing of the paper by Einstein, Podolsky and Rosen, and further, that he did not approve of the argumentation of that paper. In a response to a 1935 letter from Schrödinger to Einstein, in which Schrödinger approved of the EPR paper, Einstein said in a letter to Schrödinger (dated 19, June, 1935):* "Diese ist aus Sprachgründen von Podolsky geschreiben nach vielen Diskussionen. Es ist aber doch nicht so gut herausgekommen, was ich eigentlich wollte; sondern die Hauptsache ist sozusagen durch Gelehrsamkeit verschüttet." (Due to difficulties of language, Podolsky wrote this after many discussions. Still, it has not come out as well as I wanted it; on the contrary, the main point was so to speak, buried in rhetoric.)

    *[Einstein Archives, Jewish National and University Library, Jerusalem, Call No. 4 :1576. For further discussion see: D. Howard, 'Einstein on Locality and Separability', *Stud. Hist. Phil. Sci.* 16 (1985), p. 171.]
64. N. Bohr, 'Can Quantum Mechanical Description of Physical Reality be Considered Complete?', *Physical Review* 48 (1935), p. 696.
65. D. Bohm, 'A Suggested Interpretation of the Quantum Theory in Terms of Hidden Variables', *Physical Review* 85 (1952), pp. 166, 180.

    It is clear from several of Einstein's writings, especially

since the 1920s, when quantum mechanics appeared on the physics scene, that (both in the open literature and in his private correspondences that have very recently become available) he did not favor the hidden variable approach toward a deterministic theory of elementary matter.

Einstein's reasons for rejecting the hidden variable generalization in quantum mechanics were both specific and general. For example, in a letter he wrote in 1953 to A. Kuperman, on Bohm's early hidden variable approach to quantum mechanics (Einstein Archives, Jewish National and University Library, Jerusalem, Call No. 4 1576: 8–035), Einstein said: *(specific):*

> From a logical standpoint there is, in my opinion, no objection against Bohm's interpretation of the formalism of the present quantum theory. From a physical standpoint, however, Bohm's attempt seems to me not acceptable . . . . Where the ψ-function (in coordinate space) is not in the environment of every point approximated by a progressive wave (in contrast with a standing wave) one gets for the impulses quantities which violate the postulate that the quantum theory should contain classical mechanics as a limit case. This has to do with the fact that according to Bohm's rule, the impulses are not determined by a Fourier development but by a local formula (in coordinate space).

In the same letter, Einstein made the following *general* comment:

> I think it is not possible to get rid of the statistical character of the present quantum theory by merely adding something to the latter without changing the fundamental concepts about the whole structure. Superposition principle and statistical interpretation are inseparably bound together. If one believes that the statistical interpretation should be avoided and replaced, it seems one cannot conserve a linear Schrödinger equation which implies by its linearity the principle of superposition of states.

The reader should take note in his *(specific)* comment above, that Einstein was not, personally, supporting the validity of a model of localized particles. Rather, he was using this model of localization in wave mechanics to be close to the view of Bohm (with hidden variables) or that of the standard quantum mechanics (without hidden variables) in order to test the consistency of such views of localization of material particles in terms of wave-particle dualism.

As we have emphasized in the text, Einstein's ontological view of elementary matter was in terms of continuous, nonsingular fields, everywhere, applying to all domains of physics, from elementary particle physics to cosmology. To demonstrate this view, consider the comments that Einstein made to D. Bohm in a letter in 1953 (Einstein Archives, Jewish National and University Library, Jerusalem, Call No. 4 1576:8–053):

> When one is not starting from the correct elementary concepts, if, for example, it is not correct that reality can be described as a continuous field, then all my efforts are futile, even though the constructed laws are the greatest simplicity thinkable. [Note here that Einstein refers to "mathematical simplicity" rather than "logical simplicity", which he does adhere to.] The fact that logically simple field equations are necessarily nonlinear, and that a consistent field theory cannot permit singularities, makes it impossible, for the time being, to draw from the theory any conclusions which would allow it to be tested empirically. This does not, however, convince me of the incorrectness of the theory. It only shows our present mathematical methods are insufficient. . .

66. E. Schrödinger, 'What is an Elementary Particle?', *Smithsonian Institution Report for 1950* (Publication #4028), p. 183.
67. J. von Neumann, *Mathematical Foundations of Quantum Mechanics* (Princeton: Princeton University Press, 1955).
68. For an overview of hidden variable theories, see: F.J. Belinfante, *Survey of Hidden Variable Theories* (Oxford: Pergamon Press, 1975).
69. Einstein's belief in the principle of relativity was based on its simplicity. In a letter to L. de Broglie, just one year before his death, Einstein said: "Die gravitationsgleichungen waren *nur* auffindbar auf Grund eines rein formalen Prinzips (allgemeine Kovarianz) d.h. auf

Grund des Vetrauens auf die denkbar grösste logische Einfachheit der Naturgesetze." (The equations of gravitation were able to be discovered *only* on the basis of a purely formal principle (general covariance), that is to say, on the basis of the conviction that the laws of nature have the greatest imaginable simplicity.) Source: L. de Broglie, 'Correspondance entre Albert Einstein et Louis de Broglie', *Annales de la Fondation Louis de Broglie* 4 (1979), p. 56.

70. Galileo expressed this idea throughout his treatise: Drake, *Two Chief World Systems* (see n. 6 above).
71. A similar argument leading to the universality of the speed of light is given by: H. Minkowski, 'Space and Time', in A. Einstein and others, *The Principle of Relativity* (New York: Dover, 1923), p. 73.
72. Einstein's latest summary of his theory is in: A. Einstein, *The Meaning of Relativity,* fifth edition (Princeton: Princeton University Press, 1956). Other works expressing his ideas about the theory of relativity are: A. Einstein, 'Autobiographical Notes', in Schilpp, ed., *Albert Einstein: Philosopher-Scientist* (see n. 50 above); the following three articles reproduced in English translation in Einstein, *The Principal of Relativity,* (see n. 71 above). 'On the Electrodynamics of Moving Bodies', p. 37; 'Does the Inertia of a Body Depend on its Energy Content?', p. 69; 'The Foundations of the General Theory of Relativity', p. 111; *Ideas and Opinions* (New York: Crown Publishers, 1954).

    I have discussed Einstein's ideas from a non-mathematical view in: M. Sachs, *Ideas of the Theory of Relativity* (Jerusalem: Israel Universities Press, 1974).
73. R.V. Pound and G.A. Rebka, 'Gravitational Red Shift in Nuclear Resonance', *Physical Review Letters* 3 (1959), p. 439; R.V. Pound and J.L. Snider, 'Effect of Gravity on Nuclear Resonance', *Physical Review Letters* 13 (1964), p. 539.
74. I have discussed the subject relativistic cosmology further in: M. Sachs, *General Relativity and Matter* (Dordrecht: D. Reidel, 1982), chapter 7.
75. M. Sachs, 'A New Approach to a Theory of Fundamental Processes', *British Journal for the Philosophy of Science* 15 (1964), p. 213; 'Space, Time and Elementary Interactions in Relativity', *Physics Today* 22 (1969), p. 51.

76. Einstein appears to have changed his mind about the idea of taking seriously the implication that the Lorentz transformations denote a physical change (of length or time). In his 'Autobiographical Notes' (in Schilpp, ed., *Albert Einstein: Philosopher-Scientist,* see n. 50 above), p. 59, he said: "strictly speaking, measuring rods and clocks would have to be represented as solutions of the basic equations (objects consisting of moving atomic configurations), not, as it were, as theoretically self sufficient entities."

    I have discussed Einstein's later views in: M. Sachs, 'On Einstein's Later View of the Twin Paradox', *Foundations of Physics* 15 (1985), p. 977. I have also analyzed this problem rigorously, from the mathematical as well as the logical perspectives in: M. Sachs, 'A Resolution of the Clock Paradox', *Physics Today* 24 (1971), p. 23.
77. H. Dingle, *Science at the Crossroads* (London: Brian and O'Keefe, 1972).
78. Sachs, 'A Resolution of the Clock Paradox' (see n. 76 above).
79. A. Einstein, 'Does the Inertia of a Body Depend on its Energy Content?', in Einstein, *The Principle of Relativity* (see n. 71 above, p. 67).
80. In Einstein's 1905 paper on the relation between energy and mass ('Does the Inertia of a Body Depend on its Energy Content?', n. 79 above), he concluded as follows: "The mass of a body is a measure of its energy content." I believe that most physicists have misinterpreted Einstein's conclusion to mean; 'mass is *equivalent to* energy'. The latter statement is clearly false since, conceptually, inertial mass is quite distinct from energy. That is to say, by their very definitions, inertial mass, *per se,* is a measure of the body's resistance to a change of state of constant motion (or rest), while energy, *per se,* is a measure of the body's ability to do work. Note, for example, to refute the 'equivalence' idea between mass and energy, there are conditions under which mass is defined where energy is not. In the global domain, where one must evoke general relativity, there are no conservation laws, thus energy is not defined in a global sense. However, mass is defined globally.

It is not clear to me that indeed Einstein did mean that energy and mass are truly equivalent. His statement could have equally meant the following: given the mass of a body, m, one may then determine the total ability of that body to do work from the equation, $E = mc^2$ (i.e. in the local domain, where E is defined). However, the origin of the inertial mass of the body is an entirely different question. I have addressed this question in my research program, where I have demonstrated a source of mass, globally (in general relativity), using the general concept of the Mach principle (in Sachs, *Quantum Mechanics from General Relativity* [see n. 4 above], chapter 4.) Also, I have elaborated on the logical relation between energy and mass in: M. Sachs, 'On the Meaning of $E = mc^2$', *International Journal of Theoretical Physics* 8 (1973), p. 377.

81. Sachs, *Quantum Mechanics from General Relativity* (see n. 4 above), chapter 5.
82. J.A. Wheeler and R.P. Feynman, 'Classical Electrodynamics in Terms of Direct Interparticle Action', *Reviews of Modern Physics* 21 (1949), p. 425.
83. Sachs, *Quantum Mechanics from General Relativity* (see n. 4 above), chapter 5.
84. Sachs, *Quantum Mechanics from General Relativity* (see n. 4 above), chapter 7.
85. Sachs, *The Field Concept in Contemporary Science* (see n. 34 above), p. 69; Sachs, *General Relativity and Matter* (see n. 74 above), chapter 1.
86. Sachs, *Quantum Mechanics from General Relativity* (see n. 4 above), chapter 6.
87. For a history of the concept of complementarity, see: G. Holton, *Thematic Origins of Scientific Thought* (Cambridge: Harvard University Press, 1973). For further discussion of complementarity see: N. Bohr, *Atomic Physics and Human Knowledge* (New York: John Wiley and Sons, New York, 1958).
88. J.S. Bell, 'On the Einstein Podolsky Rosen Paradox', *Physics* 1 (1965), p. 195. Other pertinent papers on theoretical and experimental studies of Bell's inequalities are as follows: F. Selleri and G. Tarozzi, 'Quantum Mechanics Reality and Separability', *Rivisita del Nuovo Cimento* 4 (1981), p. 1; A. Garuccio and V.A. Rapisarda,

'Bell's Inequalities and the Four-Coincidence Experiment', *Nuovo Cimento* 65A (1981), p. 269; V.A. Rapisarda, 'On the Measurement by Dichotomic Analysers of the Polarization Correlation of Optical Photons Emitted in Atomic Cascade', *Lettere al Nuovo Cimento* 33 (1982), p. 437; F. Falciglia, A. Garuccio and L. Pappalardo, 'Rapisarda's Experiment: On the Four-Coincidence Experiment, A test for Nonlocality Propagation', *Lettere al Nuovo Cimento* 34 (1982), p. 1.

I have examined Bell's inequalities in the context of a theory of matter that is an asymptotic approximation for a generally covariant theory of inertia in: M. Sachs, 'Bell's Inequalities from the Field Concept in General Relativity', *Nuovo Cimento* 58A (1980), p. 1. Also see Sachs, *Quantum Mechanics from General Relativity* (see n. 4 above), sec. 2.6.

89. Schilpp, ed., *Albert Einstein: Philosopher-Scientist* (see n. 76 above), p. 59.
90. Sachs, *Quantum Mechanics from General Relativity* (see n. 4 above), chapters 6, 8.
91. P.T. Raju, 'Religion and Spiritual Values in Indian Thought', in C.A. Moore, ed., *The Indian Mind* (Honolulu: University of Hawaii Press, 1967), p. 208.
92. F.H. Cook, *Hua-Yen Buddhism* (University Park: Pennsylvania State University Press, 1977).
93. W.-T. Chan, *A Source Book in Chinese Philosophy* (Princeton: Princeton University Press, 1963).
94. M. Buber, *I and Thou* (New York: Charles Scribner's Sons, 1958).
95. M. Sachs, 'Comparison of the Field Concept of Matter in Relativity Physics and the Buddhist Idea of Nonself', *Philosophy East and West* 33 (1983), p. 395.
96. Cook, *Hua-Yen Buddhism* (see n. 92 above).
97. Buber, *I and Thou* (see n. 94 above).
98. Chan, *A Source Book in Chinese Philosophy* (see n. 95 above).
99. M. Buber, *The Way of Response* (New York: Schocken Books, 1966), p. 42.
100. For an interesting discussion of Buber's *I-Thou/I-It* complementarity, see, J. Agassi, 'The Consolation of Science', *American Philosophical Quarterly* 23 (1986), p. 129.

101. Chan, *A Source Book in Chinese Philosophy* (see n. 95 above).
102. P.T. Raju, 'Metaphysical Theories in Indian Philosophy', in Moore, ed., *The Indian Mind* (see n. 91 above).
103. For a clear statement of the epistemological differences between Bohr and Einstein, see: Schilpp, ed., *Albert Einstein: Philosopher-Scientist* (see n. 50 above), pp. 201, 666.
104. H. Franck, *The Kabbalah* (New York: Bell Publishing Company, 1940).
105. Swami Prabhavananda and C. Isherwood, trans., *Bhagavad-Gita* (New York: New American Library, 1944).
106. Franck, *The Kabbalah* (see n. 104), p. 114.
107. Plotinus, 'The One', in: D. Runes, ed., *Treasury of Philosophy* (New York: Philosophical Library, 1955), p. 960.

Index